Aerospace Predictive Maintenance: Fundamental Concepts

Aerospace Predictive Maintenance: Fundamental Concepts

SAE INTERNATIONAL

Warrendale, Pennsylvania, USA

400 Commonwealth Drive
Warrendale, PA 15096-0001 USA
E-mail: CustomerService@sae.org
Phone: 877-606-7323 (inside USA and Canada)
724-776-4970 (outside USA)
FAX: 724-776-0790

Library of Congress Catalog Number 2020947760
http://dx.doi.org/10.4271/9780768099966

ISBN-Print 978-0-7680-9427-5
ISBN-PDF 978-0-7680-9996-6

To purchase bulk quantities, please contact: SAE Customer Service

E-mail: CustomerService@sae.org
Phone: 877-606-7323 (inside USA and Canada)
724-776-4970 (outside USA)
Fax: 724-776-0790

Visit the SAE International Bookstore at books.sae.org

Chief Growth Officer
Frank Menchaca

Publisher
Sherry Dickinson Nigam

Director of Content Acquisition and Development
Monica Nogueira

Director of Content Management
Kelli Zilko

Production and Manufacturing Associate
Erin Mendicino

contents

CHAPTER 1

An Engineer's Journey

This chapter explains the author's engineering journey, where his changes in career, industries, job roles, experience, and skills have resulted in unique insights that are relevant to this book. The author came to the aerospace industry specializing in gas turbines from a background in the military marine–nuclear sector, both having safety–criticality at the core of their DNA, but subtly approaching safety in different ways.

A feature of the aerospace safety culture is a rigorous and demanding method for approving planned maintenance (PM) relating to flight safety. The official maintenance must be justified and proven to reduce measurable risk to acceptable levels. Predictive maintenance (PdM) is not yet (apart from one or two examples) accepted as mainstream PM technique, as there is much work to do to justify its use with regulators, manufacturers, operators and the maintenance repair and overhaul organizations (MROs). This book will describe ways to make PdM more mainstream. PdM is being used in most industries, but the output is mainly advisory.

Additionally, the author came from working in the in-service phase of the complex machinery lifecycle during his naval career. Having a mature in-service perspective often brings insights that even the most talented designers who have only worked in design may never realize.

One of the author's wishes is to break down barriers and improve communication in the engineering community, thus supporting Product Lifecycle Management (PLM), Integrated Vehicle Health Management (IVHM), and Systems Engineering that claim to address the whole product lifecycle. It is notable that these initiatives began their lives in the design-and-build phases of the product lifecycle, and a mistake the author perceives is that in-service phase processes and standards are being predominantly defined by design experts. One standard discussed extensively in this text is Reliability Centred Maintenance (RCM).

The author joined the Royal Navy as an Artificer (Technician) apprentice in 1974. He trained as an Ordnance-Electrical specialist but later changed designation to Marine-Engineering Electrical with post-apprenticeship qualifications in nuclear engineering in the Submarine Service.

With gained seniority, the author worked in all the specialist sections conducting defect repair, maintenance, and operation of submarine nuclear propulsion plant and ship systems. Later, as the senior electrical technician on-board, called the "Chief EA" (Electrical Artificer), the author conducted much of the preventative and corrective maintenance planning and scheduling.

The demands of nuclear submarine maintenance are exacting for a number of reasons:

- The submarine has a very wide variety of equipment and machinery in order to sustain it. As the submarine operates independently of the air, it needs to:
  - Have a powerful power plant for propulsion and power production.
  - Be habitable for extensive periods of time under water independent of the atmosphere.
  - Carry and operate weapon systems.
  - Carry and operate sophisticated sensor and communication systems.
- Additionally, in order for the submarine to remain undetected, all of this machinery has to operate making minimum noise which makes systems more complex.
- The machinery is packed into constrained spaces and so the equipment is often of compact design where accessibility may be traded off.
- Many systems store different forms and extremes of energy, making the working environment relatively hazardous. Many of the materials are hazardous.
- The number of crew is limited, and they have to be multi-skilled and highly capable.

Submarine maintenance has been equated to the complexity of space shuttle maintenance. Many maintenance tasks require contortionist skills and the dexterity of a surgeon. The part of the equipment being worked on is often out of the line of sight of the maintainer, so the maintenance work is frequently done by using mirrors and feel.

With a submarine on patrol, it is vital the crew is able to act without support. It is vital to maintain the submarine's patrol goals that are "to remain on patrol as an effective military asset, remaining undetected." Submarine systems and machinery have a high degree of redundancy which helps mitigate against the effects of failure, but this also increases the complexity of the installed systems. In the trade, submariners often describe the analogy of having a belt, braces, and string to keep their trousers up (an example of triple redundancy). Redundancy in a machinery context means many systems or machines are duplicated, and able to operate in case the first-duty machine breaks down. Standby machines or systems having interconnections (or cross-connections) to allow them to start or take over the duty of a failing machine are common. This allows submarines to live with some machinery failure, because we can use their standby machines or reconfigure the systems to restore the function. This may take the immediate pressure off of defect repair, but it does not eliminate it. It is still highly desirable to fix what is broken and restore the robustness of a fully working system.

The processes of forming Procedure Authorization Groups (PAGs) for any emergent work that had nuclear or safety implications on-board involved many of the engineering team. Great responsibility was placed on young people empowered to speak up and express concerns to ensure work was safe and effective. This ethos was rewarding and enhanced the already high degree of military teamwork. As a senior person, I relied on 17-year-old stokers working in the lower parts of the engine room to take the right actions that my own life depended on. The ingenuity and problem solving used to keep the boat on patrol as an effective military asset is an attractive part of the job.

On a submarine it is necessary to remain vigilant to maintain water-tight integrity (submarines do not have a large reserve of buoyancy). There is also a need to keep a minimum amount of systems in commission in order to maintain reactor and nuclear containment, and remove residual decay-heat from the shutdown nuclear reactor. Maintenance must also be conducted with many systems containing stored energy and where some of the plant had to be kept operational. The management of safe-to-work boundaries with highly interconnected systems for many parallel maintenance tasks, aligned with keeping systems operational, required constant management, surveillance, and ultra-strict control. A nuclear submarine only approaches the "cold metal" state where every system may be shut down and decommissioned after a year from last operating the reactor. This occurs only in a major refit situation with the submarine in a dry dock. During a normal submarine commission (4 or 5 years), this situation hardly ever occurs. After the reactor has been shut down for months, the decay heat from the nuclear reactor reduces to a level where heat may naturally dissipate through the remaining coolant in the reactor pressure vessel. In summary there are few situations that require the level of engineering excellence that submarine operations and maintenance demands.

Military engineers are both maintainers and operators. These tasks are usually split in commercial industry. Both operating and maintaining provide two perspectives on machinery that, when combined, give a greater understanding of how plant operates and fails. Having maintenance experience also helps operators deal with an overload of data that floods in when machinery trips offline. When a machine trips there is normally a cascade of alarms and warnings as other services reliant on the tripped machine are no longer operating. Being both the maintainer and operator makes for a greater appreciation of knowing what warnings and alarms are more important and need acting on. This also works the other way around: operational experience helps the maintainers' awareness of what can and cannot be done in planning the withdrawal of equipment from service and ensuring safe-to-work boundaries are not conflicting. In most other industries operators and maintainers are separate.

The sheer depth, variety, and intensity of operating and maintaining a nuclear submarine plant leaves an indelible set of memories and experience that makes the author empathic to both maintainers and operators in other industries. Although other industries' processes and contexts may be different in which machinery is used and maintained, there is enough in common that the author found he could rapidly appreciate the challenges involved.

During the latter part of his naval career, the author joined the new UK nuclear deterrent submarine class, the Vanguard Class (SSBN, Ship-Submergible-Ballistic-Missile). This class had a significant number of new digital controls, instrumentation, and monitoring systems, including the nuclear reactor control and protection. This required the technical teams after a basic introductory course to essentially teach themselves digital electronics and data processing, with different means of fault-finding and using software as new essential knowledge to maintain the equipment. Learning on the job proved to be a most powerful way of successfully dealing with this paradigm shift. Knowing how to learn is a skill that is necessary.

There is an adage about machinery faults and fault finding (or troubleshooting): "*Mechanical problems are often easy to find, but harder to fix, whereas electrical problems are often harder to find and easier to fix.*" In my experience, an electrical maintenance engineer may have extensive training and theoretical knowledge, but it takes 2–3 years of constant defect rectification work to become a competent electrical diagnostician. A competent maintainer is 2–4 times as productive compared with a trained but inexperienced person. This realization of time to competence agrees with some theories of knowledge management, and the difference between tacit and explicit knowledge that are discussed in this book.

When the author joined HMS/M Victorious during its build, he took on a personal task to look at maintenance improvement, writing many deficiency reports identifying duplicated effort between the Preventative Maintenance (PM) regime and nuclear reactor safety checks. Some of the combined PM and safety checks required multiple disturbance of systems that would have caused post-maintenance infant mortality failures.

For his troubles, the author was invited to serve his last appointment in the Navy within the Ministry of Defence (MoD) to help fix the problem. The author undertook training to become a Reliability Centred Maintenance (RCM) facilitator and reliability engineer. The RCM course was akin to an epiphany, making sense of 20 previous years of practical maintenance work and defect rectification. The author could immediately relate to what maintenance was nugatory, because the failure patterns were not compatible with the types of maintenance tasks being conducted, and viscerally understood the importance of on-condition maintenance.

In the latter part of the author's career at sea, the Navy started using handheld vibration-analysis tools, which printed recorded acceleration spectra by frequency on thermal paper. These instruments were crude by today's standards, but they were classic and effective for on-condition maintenance tasks. Some of the crew were trained to be vibration analysts on-board. They correctly diagnosed the onset of rolling element bearing failure in a number of our important rotating machines. This would have otherwise caused machine failure on patrol that were able to be changed out during our PM periods, demonstrating the efficacy of these systems.

The maintenance review project conducted in the MoD included overseeing the contracted industry team supporting submarines who did the bulk of the groundwork. The author was the only qualified Vanguard Submariner on the combined team, and his submarine plant and operating knowledge was vital to the success of the project. The joint MoD and industry team concluded the RCM-based review which resulted in a 20% reduction in PM for no loss of safety or reliability.

During the author's last appointment, he instituted visits from the MoD design and support authority to the boat's technicians just after they came off patrol. The boats signalled significant deficiency and failures in reports they had suffered as soon as they broke their patrol. When the reports were received, they were incorporated into a database (paper at the time) of the various machinery sections at the MoD design authority. These reports became the basis for reliability analysis and any subsequent redesign work. It was the responsibility of the design authority engineers to review the reports and separate them out into groups by equipment type. They could then review equipment history and using pencil-and-paper calculations determine where unacceptable reliability problems existed.

Part of the problem with this system was the time it took for report submission, analysis, and review until anything was done about them. Feedback of consolidated defects from the whole fleet from the last period was released back to all boats every quarter, but this had minimal feedback of actions being taken from the shore authorities. Many maintainers felt that it may not have been worth the effort in submitting these reports, because they had often left for new appointments before any remedial action was taken by the design authority and reported back to the fleet. There was little incentive to provide decent data.

From the design authority engineer's perspectives, many of the reports could not be acted on because of missing and low-quality data.

By visiting the technicians on the crew after they had returned from a patrol, a joint review was held of their last batch of reports. This prompted discussion between the crew and MoD team about what other data was needed and what the MoD team was going to do with the data in the support office. The results were spectacular: there was a 16-fold increase in the number of high-quality deficiency reports and an improvement in responding to reports for follow-up action. This allowed a focused effort on more effective modification work.

This visiting regime vastly improved communications between two groups of people, providing better outcomes for the Royal Navy. It also taught the author an important lesson: That accelerated improvement happens when two communities of dedicated engineers, with mutual respect and understanding, cooperate. Continued discussion between design and maintenance engineers was extremely valuable with insights being realized by both groups. A manufacturer or repair and overhaul (R&O) organization may not be able to meet pilots or line maintainers at the end of each flight sector, but there are opportunities for representative groups to meet periodically and discuss experiences around such issues as "No-Fault-Found" (NFF).

As part of the author's preparation for leaving the Royal Navy and starting a second career, I read for an honour's degree in Computer Science. I took computer science because my submarine was selected for a pilot project to develop one of the first computerized engineering-management programs. The consultants who conducted the requirements and wrote the software were arrogant and condescending to the crew. They talked over the crew in their own jargon and ridiculed our existing systems. The author determined never to be treated like that again. The engineering maintenance-software application was a dismal failure with the computerized processes taking twice as long as our existing manual system. I was able to reflect on this project as my ICT education progressed, and treat it as a case study in how not to run a technical project.

The author supplemented his bachelor's with a master's degree 10 years later in Information Systems. One of the highlights of both of my degree courses was the discovery of Systems Thinking and Systems Engineering. The Soft-Systems Methodology was especially enlightening since it showed that a different perspective besides determinism needs to be taken to solve complex (wicked) problems. I also learned that technical projects are mainly people projects. If people could not or would not use new technology, the project failed. The author has actively applied his formal education merging it into his specialist engineering maintenance domain in developing expertise in PdM. Other benefits of my education have been in data management, where learning about relational databases and the mathematical basis of SQL, from both academic and practical perspectives, have been invaluable in appreciating data integrity and quality.

After the Royal Navy, the author joined Rolls-Royce as a maintenance and reliability specialist. Rolls-Royce are the UK's design authority for the UK submarine fleet's nuclear steam raising plant. The author designed and project managed the build of a computerised Failure Recording Analysis and Corrective Action System (FRACAS). The FRACAS system framework was extended and electronically linked the reliability statistics of submarine equipment to the maintenance being conducted. Part of the issue in understanding submarine reliability (especially the new into service Vanguard Class) was the low number of boats (4) in the fleet, and dealing with low populations of failure events on most of the equipment. Weibull analysis was not viable, and the team had to adapt techniques such as Crow-AMSAA to determine likely failure rates and trends to compensate for the lack of data.

Crow-AMSAA is a means of tracking the cumulative ages between failures to show rates of failure developed by the US Army. The project team adopted it from the design reliability-growth process, used to improve new products, for in-service use where the number of failures had unacceptable variation. The FRACAS software application was later used by the MoD as the reliability baseline (then the legacy submarine fleet) to set improved contractual reliability targets for the new Royal Navy (RN) Astute class submarines, and pointing out areas where design improvements should be focused.

One interesting lesson was the careful design of the data structure that underpinned the application. We had a boat that was built of millions of components. This number of components was impractical to manage and we had to define a breakdown of equipment to a lowest level of functional significance that resulted in under 10,000 identified parts. The effort taken to get to this stage consumed over 50% of the project time and money. In hindsight, it was the right thing to do, but this was not without risk when project sponsors expected earlier tangible results. The project ran using a Rapid Development System called Dynamic Systems Development Method (DSDM) that was very new at that time (1997–8). The level of trust in the project was high because of the direct engagement and participation of the users, who bought into why we had to initially focus on the data design. Active involvement of influential users in software projects is very beneficial.

The author then transferred to a new Rolls-Royce joint venture company (Data Systems & Solutions or DS&S), a new group tasked to develop systems to gain access to Rolls-Royce asset usage data to de-risk Rolls-Royce power-by-the-hour™ service contracts. The initial thinking was that data access from the control systems was going to be possible, but this was subsequently not proven because of customers' sensitivity and costs for providing data without any apparent benefit. The most effective way of accessing and learning from asset data was to provide PdM services to the customers. In this way the customer attached more value to their business and provided permission to access data. In the future, data owners will realize the value of their data. In order to induce data owners to share that data, they need to realise a business benefit, and their intellectual property and competitive differentiation must be kept safe.

DS&S started with gas turbine engine health monitoring software that was given to Rolls-Royce customers after their purchase of gas turbines. It is striking and remarkable that a business was started based on software that was given away free. The new company expanded this to offering fee-paying services for in-house analysis and alerting using the software. The original value proposition was based on DS&S expertise of the health monitoring software to interpret the interim results and summarize them. Later additional benefit was realized because the size of the fleet being monitored from many customers enabled the service to learn lessons from one customer that could be applied to all. This service was very successful, eventually covering the whole range of Rolls-Royce's and other manufacturers' products. The health management project started just before the dot-com boom in the early 2000s where the service team was posting customer-analysis results in pdf documents for Engine Health Monitoring (EHM) on the web.

The author was a key member of the team leading Research and Development (R&D) and became the corporate owner for the PdM capability over many years, before Big Data, applied Artificial Intelligence (AI), machine learning (ML), and predictive technologies emerged. The Rolls-Royce system was redeveloped through several generations over the years and is still at the forefront of PdM capability globally in any industry sector. The Rolls-Royce PdM system adds hundreds of millions of dollars of value annually to airline operators, Maintenance Repair & Overhaul (MRO) shops, and Rolls-Royce itself.

During this time the author participated in aerospace and oil and gas maintenance standards writing, becoming team leader in describing PdM for API 691, Risk-Based Machinery Management (for rotating machinery) and contributed several chapters to SAE books on Integrated Vehicle Health Management (IVHM), and other books on Through Life Engineering Services. The author is co-inventor of a PdM patent in simplifying diagnostics.

In late 2014, the author took voluntary redundancy from Rolls-Royce and soon after co-founded Ox-Mountain Ltd. Ox-Mountain successfully developed service software that applies ML to machinery data to automate engineering and reliability processes in the process industries sector (mainly mining). Being part of a small team and having to turn his hand to various diverse tasks, the author discovered the power of the programming language Python and spent the last few years learning how to prototype solutions using it to gain insights from the data.

Who Is This Book Intended For?

The goal of this book is to explain PdM in a simple and understandable way. The intent is to de-mystify core concepts and challenge hype. It is also intended for students, technical managers, non-engineering managers who may rely on or are involved with financing asset maintenance. This book is intended to explain PdM in an approachable way to enable those people to make more informed decisions in adopting or using the technology.

PdM is a relatively new capability and has not reached full maturity. It is still subject to a high degree of marketing hype and claims attempting to differentiate many suppliers' sales pitches. This leads to over-exaggeration and false claims that result in confusion and misunderstanding that the author hopes to clarify.

This book posits that PdM is a subset of Condition-Based Maintenance (CBM) and must obey the same underlying rules and pre-requisites that apply to CBM. CBM is further explained in the Glossary. PdM is new because it takes advantage of newly emerged digital technology, in sensing, acquiring data, communicating the data, and processing it but it is a subset of CBM. The more recent advances in Big Data and predictive technologies has also accelerated PdM development.

Applying digital technology allowed sensors to be continuously monitored, and timeliness of processing is also enhanced, widening the ability to take practical remedial action if a failure is detected. Other major differences between traditional CBM and PdM are:

- The fidelity of on-line sensing can be lower compared with CBM periodic sampling using handheld equipment. This is because taking one sample requires that the machine needs to operate without failure for a considerable period longer than being able to derive readings from online systems much more frequently.
- The data processing and analysis lifecycle may be highly automated, with applied ML and AI (predictive technologies). This capability can autonomously analyze the data and send alerts and advice to decision makers, potentially reducing through life cost and improving safety.
- The high automation inherent in a quality PdM system also boosts scalability, the PdM system is capable of dealing with many more assets' health compared with manual or immature PdM systems that require a high degree of human intervention.

FIGURE 1.1 The acceleration of emerging technology adoption (Source New York Times).

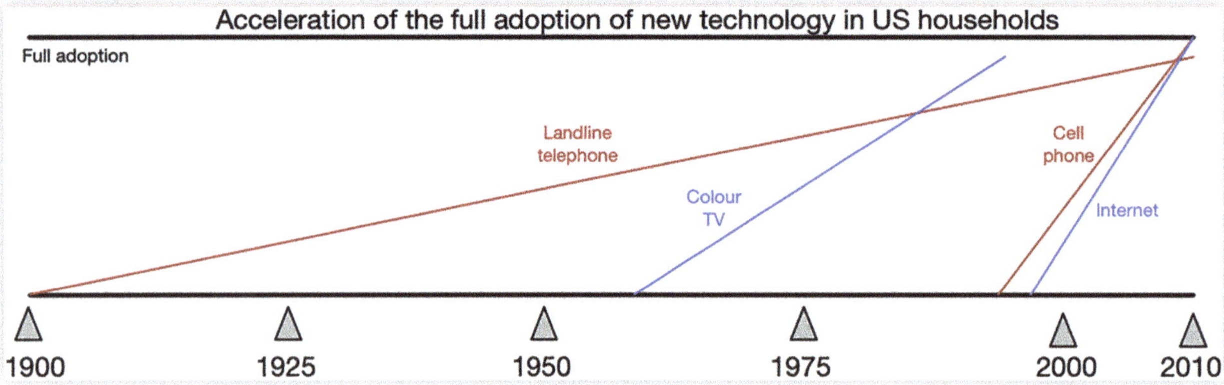

In the wider world the rate of technology innovation and adoption is accelerating, illustrated in Figure 1.1. This is the case with the development of PdM. The internet means that adoption times can be very short, given its global reach and connection to most people on the planet. Think how fast automated security software patches are adopted for computer operating systems to understand how fast adoption could be. This title will explore how PdM may evolve, how it fits with other emerging technological development, and how it may take advantage of it. It provides a context for how PdM may be further developed in the future. This book does not cover an explanation of some of the underlying on condition techniques such as vibration analysis, these techniques are explained in detail in another complementary book by Keith Mobley (Chapter 1 reference 1).

PdM has been applicable to any industry that relies on complex physical assets that require maintenance. As PdM has emerged and has been maturing, especially with automation, the scope of its applicability is widening. PdM can be applied to smaller lower value assets where sensor and other data is available. This trend of ever wider applicability will continue and organizations' operational effectiveness and efficiency may be increased by small increments that feed directly to their bottom lines. Examples will be drawn from many industries, but the book recognizes that the aerospace industry is most likely to be the leading user and developer of the technology. There are other industry sectors that have distinct requirements where PdM and machinery monitoring they use are specialized and world leading. For example, oil exploration where directly monitoring and controlling the drilling bit is paramount. This book has an aerospace focus but should be directly relevant to other industry sectors.

How This Book Is Organized

Many history texts that describe a specific but highly influential event often set the context by describing the preceding story in detail. This prehistory helps the readers comprehend why the events unfolded, deepening the appreciation of how and why the central subject was so influential. It is so in this book. Some of the chapters describe other, but related technologies and frameworks that might at first glance seem disconnected. This risks possible inconsistency and confusion for the reader. This section will provide an overview of how the chapters fit together, what relationship and influence to PdM they have and how they all help to gain both a broader and deeper appreciation of the technology.

Chapter 1 describes the author's unusual career path to establish a frame of reference for his worldview and perspectives. The chapter provides the reader with an insight for the author's frame of reference and how this influences his views of PdM. This allows the reader to reflect on their own frame of reference and empathise or disagree with the content of this book. If any reader has contrary views, the author would welcome any constructive feedback with reasoning.

Chapter 2 provides a history of maintenance, how over millennia it has developed and adapted in response to growing complexity of machinery over time. Not only has machinery complexity increased, but what society demands of modern machinery has considerably increased in scope as well. Performance, safety, and the environment make direct demands on maintenance to help ensure these requirements are met and managed. With increased demand, maintenance has to deliver more for less. An important story told in Chapter 2 is the generational thinking about the patterns of failure that exist. Many people are taught about the second-generation model, called the bath-tub curve, but do not realize that a third generation exists subsuming the older model produced after extensive research with empirical evidence. The third generation of failure patterns are vital to understand why on-condition and PdM are so important.

Chapter 3 initially introduces maintenance and the purpose of maintenance from a high-level perspective to set a context of where PdM sits and operates in the wider maintenance regimes. PdM is not a panacea, and does not make other maintenance obsolete. PdM is a component of Preventative Maintenance that preserves functions (preventing machinery functional failures). The chapter broaches functional failure and what aspects in the design, manufacture, operating, and environmental contexts influence failures and failure rates. It shows that much of the background knowledge around failure can be contained and structured in a Failure Modes and Effects Analysis (FMEA). A Taxonomy of maintenance task types is provided. The chapter then describes the core PdM in detail, showing conceptual models of what types of failure are applicable for PdM, the concept of a model of normality, novelty, and anomaly detection, how PdM breaks down into diagnostics and prognostics before alerting those that need to know. The chapter describes some of the hype surrounding PdM, before finally showing how it impacts and disrupts traditional maintenance planning and scheduling.

Chapter 4 covers Integrated Vehicle Health Management (IVHM) that aims to provide a platform-centric framework for PdM. IVHM is not covered in detail but enough information is given for the reader to find further information. IVHM is a whole product lifecycle approach that uses Systems Engineering to conceptualize design, build, and operate PdM for any complex asset. The aerospace sector has a growing number of Aerospace Recommended Practices (ARPs) for IVHM including PdM. In the aerospace sector, formal maintenance is defined using the MSG3 RCM methods with enough evidence to formally assure flight safety. PdM has yet to be adopted in this formal process, but the IVHM, SAE, and RaeSoc and industry regulators are working toward agreed processes to formally adopt PdM techniques so they may deliver maintenance credits, thereby reducing risk of failure to acceptable levels.

Chapter 5 explains the economic advantages of applying PdM instead of other traditional fixed-interval preventative maintenance, where it is practical to do so. This is an important benefit to be applied when building a business case for adopting PdM. The explanation introduces the Weibull distribution and its associated cumulative density function to illustrate the economic advantage of PdM over scheduled replacement or restoration tasks. The chapter points out that the maximum potential value for PdM would be the utilization of components for a population that would be the median value of the Weibull characteristic, in addition to avoidance of disruption. The chapter describes how RAM simulation (using Weibull distributions and Monte Carlo techniques) may be used to provide the data for business cases.

Chapter 6 takes a closer look at how PdM relates to RCM (or MSG-3). It highlights a common misunderstanding that PdM is superior or supersedes RCM, and explains why this is nonsense. RCM is used to design and specify maintenance, PdM is an important constituent of maintenance. This chapter shows what other data should be included in a FMEA to make it better for RCM, but also to provide vital information to judge whether PdM is applicable or not. The chapter provides a light introduction to a graph structure, and then shows a schema for a FMEA drawn up in Figure 6.2. The chapter moves onto the second stage of RCM where a decision logic is used to assign different maintenance tasks to the underlying failure modes and shows the criteria used to make the choices.

Chapter 7 discusses PdM maturity. To provide a context, the transformation of data through information and knowledge is described, because PdM is massively data dependent as a system. Understanding some of the precepts of knowledge management provides a really useful and powerful perspective on PdM as an information system. A critique is provided for a traditional knowledge management system, with an improved version that the author puts forward that helps conceptualize PdM. The chapter goes on to discuss Data Quality to ensure that both hard and soft attributes of quality are included. The chapter moves on to provide a functional breakdown of a PdM system used by IVHM, called SATAA. This book extends the model to include Learning (SATAAL) to illustrate that an operational information system needs feedback and reflection to continuously improve. A PdM system needs to adapt as assets age, new technology emerges and improve as lessons are learned. SATAAL is described in detail, with salient points explained in more detail that illustrates what needs to be in a PdM system. In order to operate a successful PdM system, a number of necessary competencies and roles are put forward. Finally, a simple five level maturity model is described that show major elements of what have been described in the chapter are ranked.

Chapter 8 provides a template functional specification for a PdM system, showing importance of the requirements. The requirements are based on discussions in the earlier chapters and may be used as a datum set of requirements for any organization that wants to specify or assess a PdM system.

Chapter 9 discusses disadvantages of PdM and shows how these may be addressed. One of the fundamental changes PdM implies is a shift from deterministic black and white thinking to more nuanced decision making informed by probabilities and uncertainty. This was discussed in building a business case but is also vital in how the recommendations from a PdM system are trusted and acted on. PdM implies that future forecasting of maintenance events is more uncertain compared with a traditional time-scheduled based system. This may be mitigated by delivering a usefully long P-F interval to enable planning, predisposition of resources and timely withdrawal from service minimising disruption. Other concerns such as data management, privacy, and ownership are discussed.

Chapter 10 is a forward-looking view of PdM. At the very beginning of the book we illustrate the ever-accelerating emergence of new technology, and this is almost certain to influence the further development of PdM. We describe Big Data and Cloud which is already being widely adopted, but also cover newly emerging technologies such as the Industrial Internet of Things (IIoT) that will result in proliferation of cheap, wireless, ultra-low-power sensors with a tsunami of increased data volumes. IIoT will transform PdM, because it will make so much more equipment and their environments economically monitorable. Industry 4.0 is covered, which exploits IIoT, and how the data produced in manufacturing should be exploited by PdM. The sharing of data between different phases of the equipment lifecycle (between design manufacturing and in-service support) may be improved through Industry-4.0 adoption. A more blue-sky technology is discussed using nanotechnology that can be used for new sensors, micro-robotics for inspections and self-healing or repairing systems that may be integrated with PdM.

Chapter 11 provides a summary, but also offers some indicators and pitfalls to avoid when implementing a PdM system. How you might start and rapidly scale up after initial value has been demonstrated. How knowledge is being lost due to demographics and how this might be compensated. This ties back to an appreciation of knowledge management covered in chapter 7. A couple of described examples of how PdM works provide a "what does PdM feel like" when it is adopted.

Chapter 12 shows an example of a signal processing using a single variable Kalman Filter to smooth a time-series data trend. The filter is coded in Python and run to show some examples of the smoothing using example data. The Kalman filter is shown compared with using other moving average algorithms. The example also shows how the Kalman filter may also be exploited to deliver "novelty" or "anomaly" detection, where trends either exceed a threshold, step change or ramp, so that these can be further used in the Sense - Acquire - Transfer - Analyse - Act - Learn (SATAAL) Analysis stage as inputs to diagnosis.

A Glossary is provided at the end of the book.

Figure 1.2 shows the book chapters and how they fall into three groupings to describe the context, then the PdM technology followed by a small case study and what advances in the future may occur.

FIGURE 1.2 An overview of chapters and flow through this book.

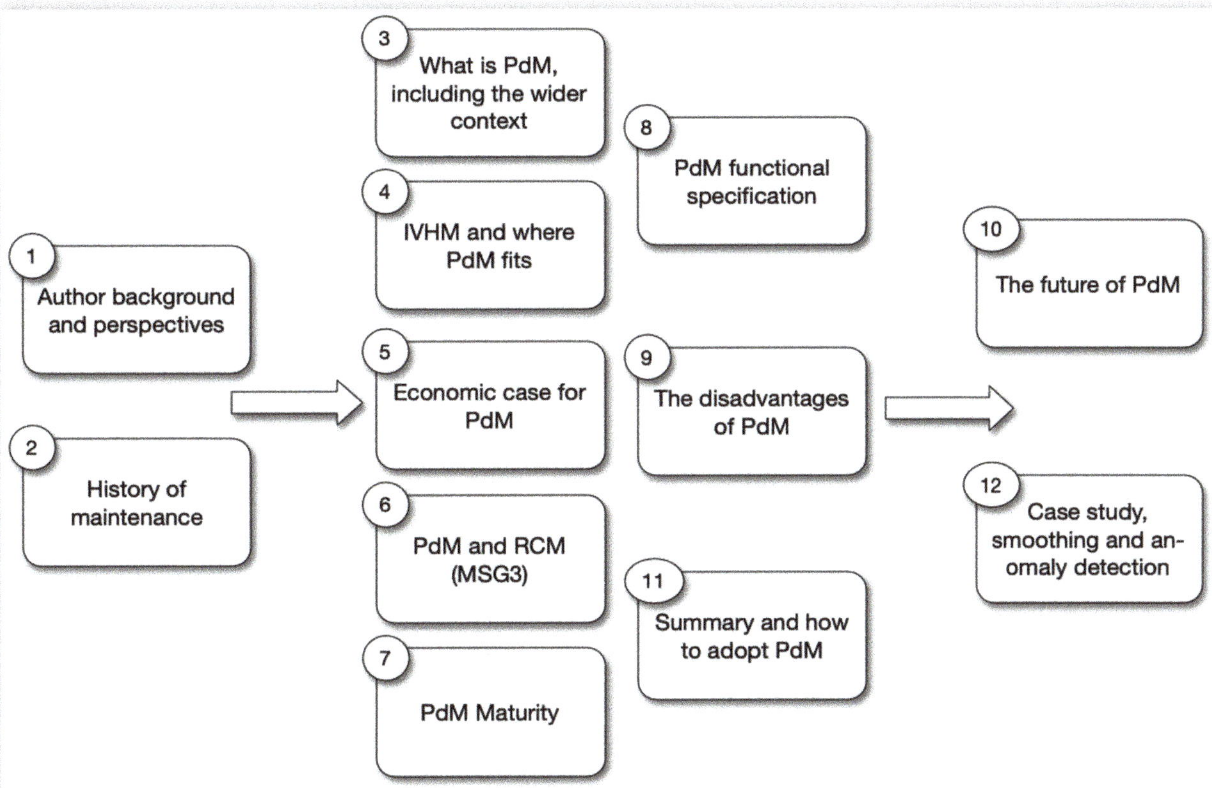

Reference

1. Mobley, R.K., *An Introduction to Predictive Maintenance*, 2nd ed. (Butterworth Heinemann, Amsterdam London, New York, Oxford Paris, Tokyo, Boston, San Diego, San Francisco, Singapore and Sydney, 2002).

CHAPTER 2

A History of Maintenance and How Maintenance Is Done Today

To set the scene for PdM, the history of maintenance will be discussed through millennia up to modern times, including the more recent developments in aerospace with the "Air Transport Association" (ATA), now called the "Airlines for America" (A4A), MSG-3 Recommendations, and the adoption of Reliability Centred Maintenance (RCM) in other industries. Other standards are also emerging for many industries that take a wider perspective of Asset Management, such as the ISO 55,000 series [1]. These other standards embrace maintenance and maintenance management and use many of the systems and frameworks described in this book.

Maintenance has been conducted for thousands of years in distinct phases as the complexity of machinery and technology has evolved. The first phase of maintenance was characterized by simple wooden machines, where breakage was repaired by using many of the natural resources in the immediate vicinity of the maintainer (or artificer). Some of the maintainers were skilled masons, carpenters, or blacksmiths, but they were relatively independent and self-sufficient. Some of the most sophisticated devices built up to the industrial revolution were clocks and timepieces, especially those built for maritime navigation in the 18th century.

This continued until the Industrial Revolution. The British Royal Naval dockyards invented the earliest forms of industrial manufacturing during the Napoleonic wars, producing standardized cordage, rigging and other replacement parts using proto-production lines to keep the vast fleet of the Royal Navy operating. I. K. Brunel's father, Marc Brunel*, was responsible for building a series of jigs for making rigging blocks used to improve manufacturing efficiency by 11-fold [2] in the Royal dockyards.

When steam power was introduced, machinery took a step-change increase in complexity and introduced the second phase of maintenance. Although the majority of

* The Brunels were famous Victorian engineers who both set new engineering records and had similar reputations and fame in their time as Elon Musk has today.

the maintenance was corrective (fixing equipment after it has broken), some servicing (mainly lubrication and cleaning) and restoration (restoring the condition of the machine to nearly new) maintenance was introduced to restore condition or performance. For example, boiler tube cleaning was periodically conducted to bring back heat transfer performance and efficiency of steam production. The predominant belief in how equipment failed at this time was due to wear-out, where the probability of failure increased with usage of the asset or its time exposed to the environment. A summary timeline of major innovations and inventions during the Industrial Revolution is shown below (Figure 2.1).

Machinery complexity took another step change in the 1930s with the introduction of monocoque multi-engine aircraft. The sophistication of aircraft was also accelerated by WWII. When you can compare a Boeing B29 Superfortress from 1944 with a pressurized heated cabin, centrally directed powered guns, new avionics and hydraulics to another simpler aircraft like a Boulton & Paul Overstrand, it is easy to see at least two generations of development between these two aircraft types. Jet engines also made their operational appearance during WWII (Figure 2.2).

Although these newer aircraft systems were complicated, their reliability was relatively poor compared with today's standards. Typical serviceability figures for the RAF in the 1940s varies between 60% and 80% availability. The third generation of maintenance was initiated by the introduction of preventative maintenance with scheduled overhauls and replacement of parts after a period of time before they wore out. The widely held belief at the time was that the more a complex machine was maintained the better its reliability.

The introduction of jet engines, pressurized cabins, and the discovery and fixing of low-cycle fatigue failure modes with the Britannia Comet aircraft initiated mass travel. The complexity of aircraft systems increased with the introduction of solid-state electronics (Figure 2.3).

The beliefs in reliability morphed to the bathtub model where there may be a stage of infant mortality failure before a period of useful life prior to wear-out. Concepts such as "burn-in" were adopted to run components on load just after they were manufactured to reduce the incidence of in-service failure due to infant mortality, especially in

FIGURE 2.1 Major innovations in the first industrial revolution.

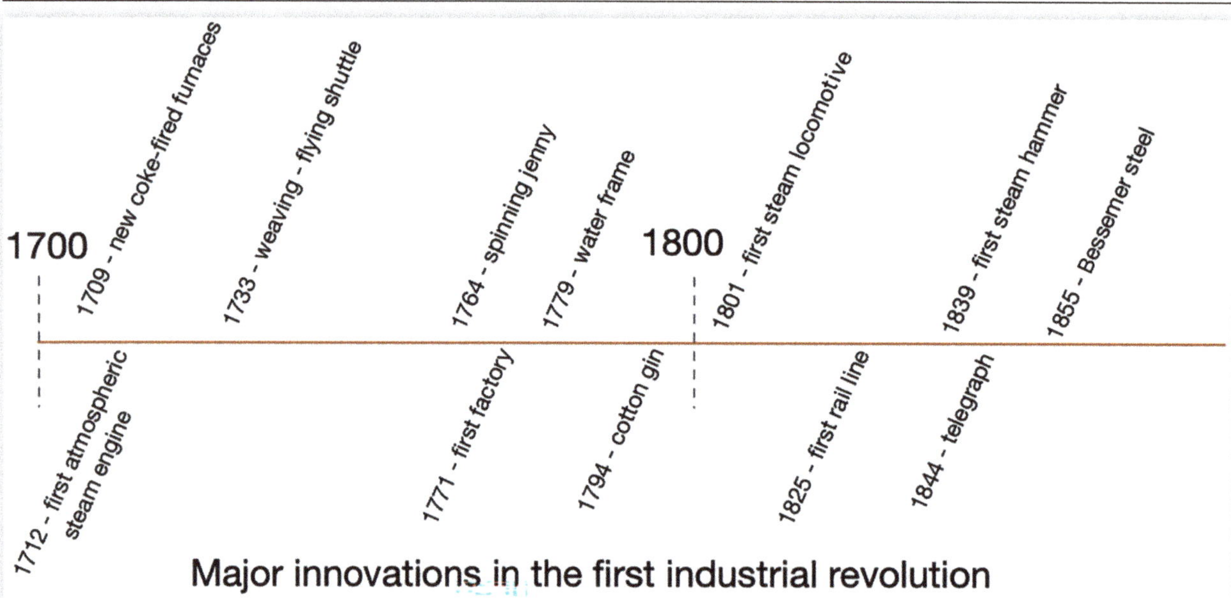

FIGURE 2.2 Comparison of the complexity of aircraft spanning WWII (Source Wikipedia).

FIGURE 2.3 The old superseded bathtub belief of how machinery fails.

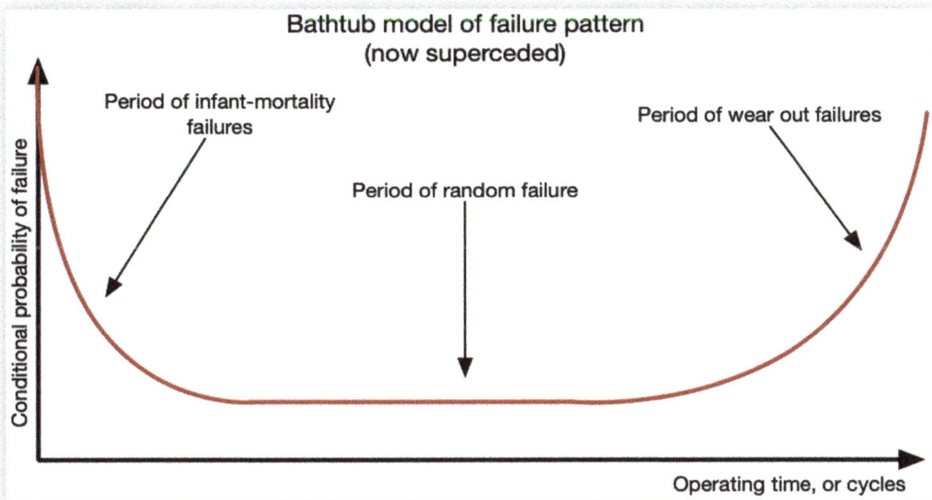

electronics. A major misunderstanding about infant mortality is that a period of infant mortality may be initiated after invasive planned maintenance has been conducted, where the work quality may be substandard. This complicates the simplistic bathtub characteristic as it may not be uniform over the full in-service phase of the product lifecycle (Figure 2.4).

In the late 1960s, The US Aerospace Maintenance Steering Group (MSG) introduced MSG1 that incorporated the concepts of engineering discipline for design safety. Overhaul and on-condition tasks were introduced to make the through-life maintenance costs of the Boeing 747 (Jumbo Jet) low enough to make the aircraft economically viable. Mass travel by aeroplanes would never have been possible without the adoption of the new MSG maintenance philosophies.

In the early 1970s, MSG-2 was released, introducing the concept of condition monitoring in time for the introduction of the Lockheed 1011 and McDonald Douglas DC-10. Equivalent standards were also adopted in Europe for Concorde and the Airbus A300.

In the late 1970s, the as-then Air Transport Association of America (ATA) instituted a review of the MSG-2 standards because of the effects of inflation with huge rises in

FIGURE 2.4 Major innovations in Aviation.

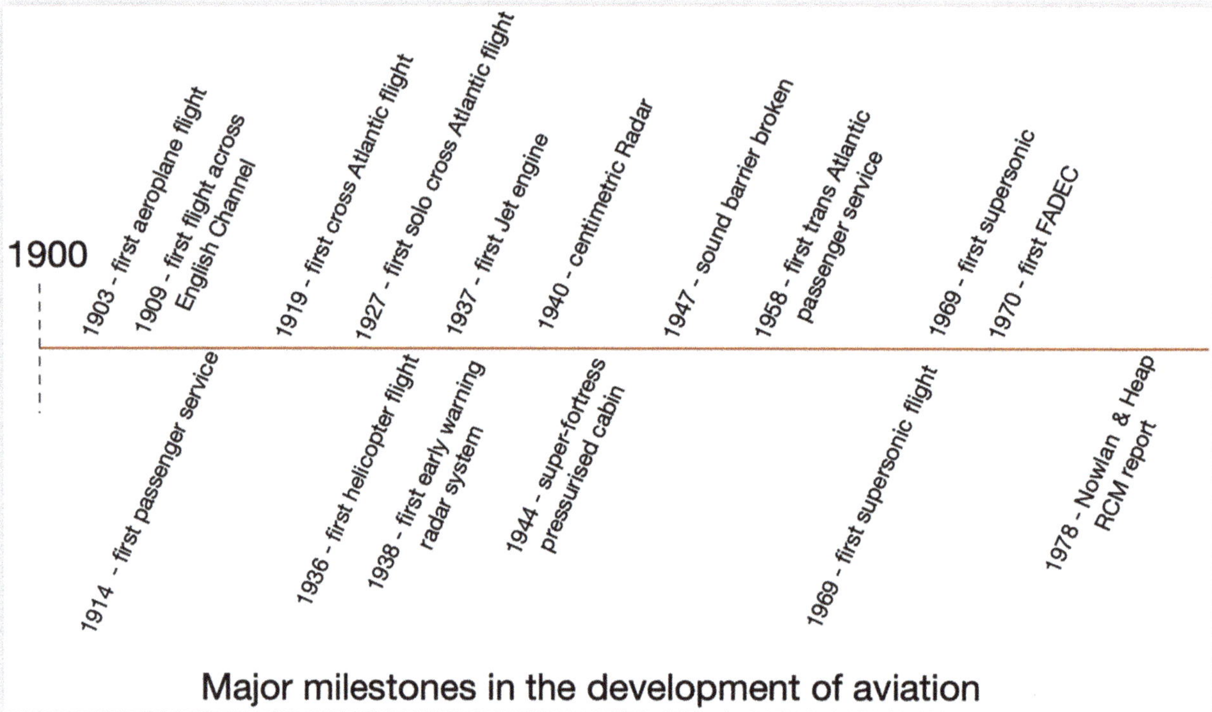

fuel costs, the introduction of other new aircraft and engines (GE CFM56), and the release of the United Airlines Reliability Centred Maintenance (RCM) report [3].

The Nowlan & Heap RCM report was seminal. It included a major empirical study that identified functional failures which fell into six distinct failure patterns, some of these being wear related, some random, and another conforming to the bathtub and infant mortality. This was a major update in the previous generational belief in the bathtub curve representing failure patterns. The study also showed that the majority of functional failures for complex machinery was random. These unexpected results were confirmed later by further independent empirical RCM studies in the US Navy and US nuclear industry. These results seem unintuitive to many engineers and may require some explanation. The following diagram shows the progression of beliefs in how assets failed over the three generations of maintenance development (Figure 2.5).

Each of the graphs above show the conditional probability of instant failure (the Y axis) against the age of the component (X-axis). The unexpected result is the predominance of random failure, which needs to be explained. Many people believe that wear-out and age-related failure predominates, dismissing the results of the RCM studies.

Many reliability and maintenance textbooks still quote the bathtub model today as being the underlying model of failure that is still prevalent. This may be due to the extra complications of explaining the six models, and the unintuitive conclusions that most functional failures in complex machinery are random. The RCM theory is not yet pervasive, except in those industries mentioned above where safety is critical. Another issue is the way reliability engineering is taught, especially if it is only a module taken as part of a general engineering degree, is that teaching the bathtub as a model for failure is far simpler than the six RCM failure patterns. The bathtub is not a satisfactory model for all failure patterns, it certainly does not represent the pattern of failure caused by fatigue.

FIGURE 2.5 Generations of belief of the patterns of failure (Source Ch 2) [4].

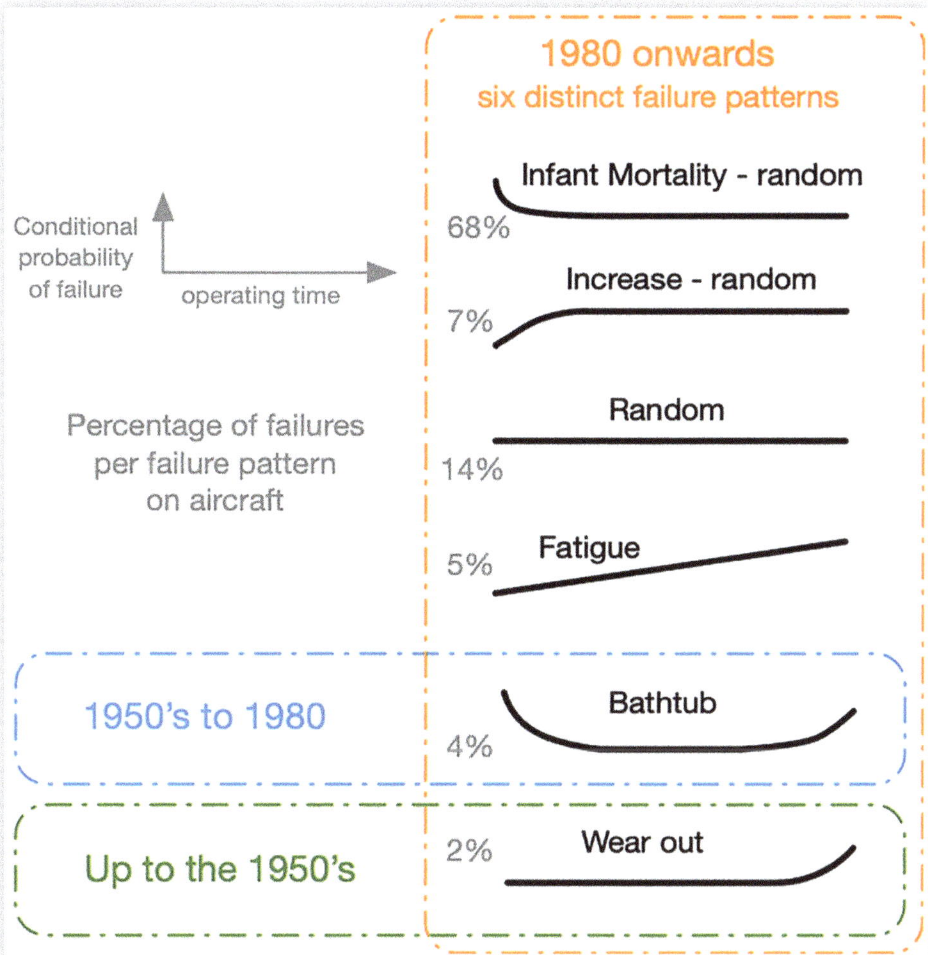

Most Functional Failures in Complex Machinery Are Random

The Nowlan and Heap study considered functional failure which may be caused by many failure modes on many components. System complexity is substantially increased by the inclusion of software.

The failure patterns are for complex machinery where the top-down approach of looking at system failure down towards components probably mixes many failure modes. Where the breakdown has occurred at individual component level individual components, such as turbine blades, the failure patterns for these parts is wear-out with thermal low-cycle fatigue, a major life-limiting factor. Other failures can be caused by environmentally driven events [such as Foreign Object Damage (FOD), including birdstrike], which are random (sometimes these are called "acts of God"). The general idea seems to be: if the part has no moving components or is a physical (structural) entity in contact with the operating medium (e.g., a turbine blade in the gas path, or a pump impellor in contact with the liquid being pumped), then the failure pattern for the predominant failure modes of that part is likely to be age or wear-out related. A structural part in contact with the operating medium is not inherently complex.

Another factor that may support the predominance of random failure patterns is that the root causes of the majority of failures can eventually be traced to human factors, such as lack of communication or incorrect decisions. These human factors are explored in Chapter 7 where Knowledge Management is covered.

Infant mortality failures (where failure occurs at unexpectedly low age) are mainly caused by quality issues in design, manufacturing, assembly, operations, or maintenance. During the in-service support period, infant mortality may still be present due to maloperation and bad maintenance. This means that infant mortality might not just be a factor at the start of the working life of a component; it is liable to recur after each maintenance period, if the maintenance procedures have quality issues. A primary function of a PdM system should be to present an analysis of failure ages of components after Preventative Maintenance to identify infant mortality. If parts are replaced with new as scheduled replacements, and the new parts are badly fitted, the Weibull analysis will also pick up maintenance-induced infant mortality.

Maloperation may also be monitored using PdM systems, showing where machinery is operated beyond its intended design boundaries. Type 2 prognostics (explained in chapter 3) can be used to update the likelihood of failure, given that damage has accumulated by maloperation. Where maloperation is due to operators, the exposure should be treated as a learning and improvement opportunity instead of a blame culture and applying a "stick" to discipline operators. Both the nuclear and aerospace industries have an open culture where operational mistake reporting is actively encouraged and shared, and treated as learning and improvement opportunities.

When conducting RCM (or MSG-3, described in detail later in the book) analysis, analysts drill down from the asset through its systems to components (known as indenture levels), identifying the failure modes (and often the mechanisms) of failure. An assessment of the impact and likelihood of the failures occurring are recorded simultaneously. As a granular level of components is reached, the failure mechanisms and the failure patterns tend toward being aged and worn-out. If functions and functional failure at the higher levels of indenture are considered, functions may fail for different failure modes and environmental effects. This implies the mixed failure modes leading to a single functional failure may appear to be random. This effect may be mitigated if all failure modes have wear-out patterns where the increases in probability of failure are varied, implying one or two failure modes are predominant, which mask the other failure modes from presenting themselves.

We need to understand that failure may be represented as a hierarchy: at the top there are functional failures, caused by one or many failure modes, each failure mode degrades because of the underlying failure mechanisms. A common mistake is to mix these entities up.

What Is the Difference between Failure Modes and Failure Mechanisms?

This explanation is included here because many people do not differentiate between Failure modes or mechanisms and frequently mix them up. One of the important things to get right in the RCM process is to ensure these are differentiated and correctly identified.

A failure mode in the RCM domain is a description of the state an object is in when functionally failed. If a machine is "leaking" at an unacceptable rate, this is describing the failure mode. It is well to remember that many machines with a containment function have an acceptable or design leakage rate. "Leaking" may be more formally termed "loss of containment." The failure mechanism leading to the leak describes the physical process by which the failure mode has occurred. The failure mechanism may be due to corrosion, erosion, fatigue, etc. All of these are plausible failure mechanisms leading to functional failures. Depending on what mechanisms are likely or predominant, they may influence choices on what sensors should be fitted to provide, such as early warning of functional failure using PdM in a particular operating context. Failure mechanisms may also work interactively (e.g., corrosion is combined with erosion). The rate of degradation is likely to increase if two or more failure mechanisms interact concurrently. The interaction of failure mechanisms is also a demonstration of holism: "the interaction is greater than the sum of the failure mechanisms acting alone."

In summary, functional failure that can be caused by a multitude of failure modes tends toward being random. Complex machinery and components (with software) may tend toward random. Simpler, more structural parts or components with few failure modes tend toward presenting wear-out and age-related behavior.

Intrinsic and Achieved Reliability

Many people assume that improving maintenance leads to ever improving reliability, this is a mistake. Maintenance (including PdM) cannot exceed intrinsic reliability set as an upper level of achievable reliability by design and manufacturing of equipment. However, it is possible that the reliability achieved in service may be far less than intrinsic reliability due to maloperation and mal-maintenance. The in-service efforts in operations and maintenance should aim to attain levels equal or close to intrinsic design reliability. Only design change, or modifications can change machinery intrinsic reliability.

Applied Systems Thinking

The author is a convinced adherent of "Systems Thinking" and how this has been applied in the engineering domain of "Systems Engineering." It is valuable to apply systems thinking to equipment to determine the best way a project team can preserve its functions in the most effective and efficient way. This section will describe reductionist and systems thinking ideas and how this applies to analysing equipment to determine how it may be maintained, including using predictive maintenance.

During the last 300 years, reductionist thinking was used to divide problems into smaller components, so better understanding about the smaller constituents is possible to resolve the larger problem. Reductionism remains a very potent strategy for problem solving and needs to be used in conjunction with systems approaches. Systems thinking and systems engineering does not reject reductionist processes but augments them as described below.

As equipment has increased in complexity over time, projects that have relied on reductionist approaches have increasingly failed. A new approach to address complexity and the limitations of reductive thinking was the development of Systems Thinking and Systems Engineering.

Systems Thinking adopts the concept of a system that has a boundary and exists in an environment that it interacts with. The systems approach embraces Reductionism but adds the concept of Holism; when systems are observed as an entire entity, they have properties that are not present if you observe all isolated system components. This idea of Holism is best encapsulated when Aristotle said, "The whole is greater than the sum of its parts."

Other system concepts include "emergent behavior" and "self-organizing systems." Emergent behavior is when a complex system displays unexpected behaviors if system influences change or more components are added to the system. On top of emergent behavior is the systems concept of "self-organizing systems," where systems may spontaneously reorganize themselves. This phenomenon is especially true if there are associated social constituents of a system. The inclusion of people in a system requiring human intervention to operate manage and maintain the system, makes it the most complex possible and most likely to self-organize in ways that are unexpected and difficult to predict. It is often known as the "law of unexpected consequences."

The act of decomposing assets (that are systems with identified boundaries) into smaller systems to simplify and address complexity is reductionism at work. The analyst also needs to consider that a whole system may have functions and behaviors that are not apparent by analysing their parts in isolation. So, the analysis must not be restricted to a pure reductionist approach.

MSG-3 Overview

Based on the RCM principles derived from the Nowlan and Heap RCM report, MSG-3 was issued in 1980 and is the current aerospace guidance for defining maintenance regime for the manufacturers and airline operators for each type of aircraft. MSG-3 has been updated through the years with the latest version being issued in 2018. It is noteworthy that MSG-3 has remained extant for 40 years, a testament to its technical soundness based on solid empirical evidence.

MSG-1, 2, and 3 recommendations were important turning points in maintenance history and important to the invention of PdM because they introduced the concepts of on-condition maintenance, which PdM is part of. Without RCM and MSG-3 the airline industry would be uneconomic to run because maintenance costs would have been too high. The reductions in costs of flying while maintaining the rigorous standards to preserve and help improve safety derived via RCM meant the whole era of mass travel was enabled. Many people who criticise RCM as being too onerous would do well to remember how our ability to travel cheaply relies on RCM in its MSG-3 guise.

The MSG-3 standards form recommended practices that guide operators and manufacturers for the construction of maintenance programs for aerospace. The standards are split into independent sections covering fixed wing, rotary, structural systems, and powerplant and since continued to update. The latest 2018 version is available from A4A publications [5].

Figure 2.6 shows a simplified representation of the MSG-3 decision logic used to determine what maintenance task types are applicable. Starting at the top, the question of "evident" means "the failure is evident to the asset operators or maintainers when they are undertaking their normal duties." If the failure is not evident, the user of the decision logic follows the "hidden" branch. The next question asks, "does the failure have any safety impact or consequences," because then the stringency for the selected tasks will be tighter, and the burden of proof that the maintenance tasks will reduce the

FIGURE 2.6 Highly simplified MSG-3 decision logic (Source MSG).

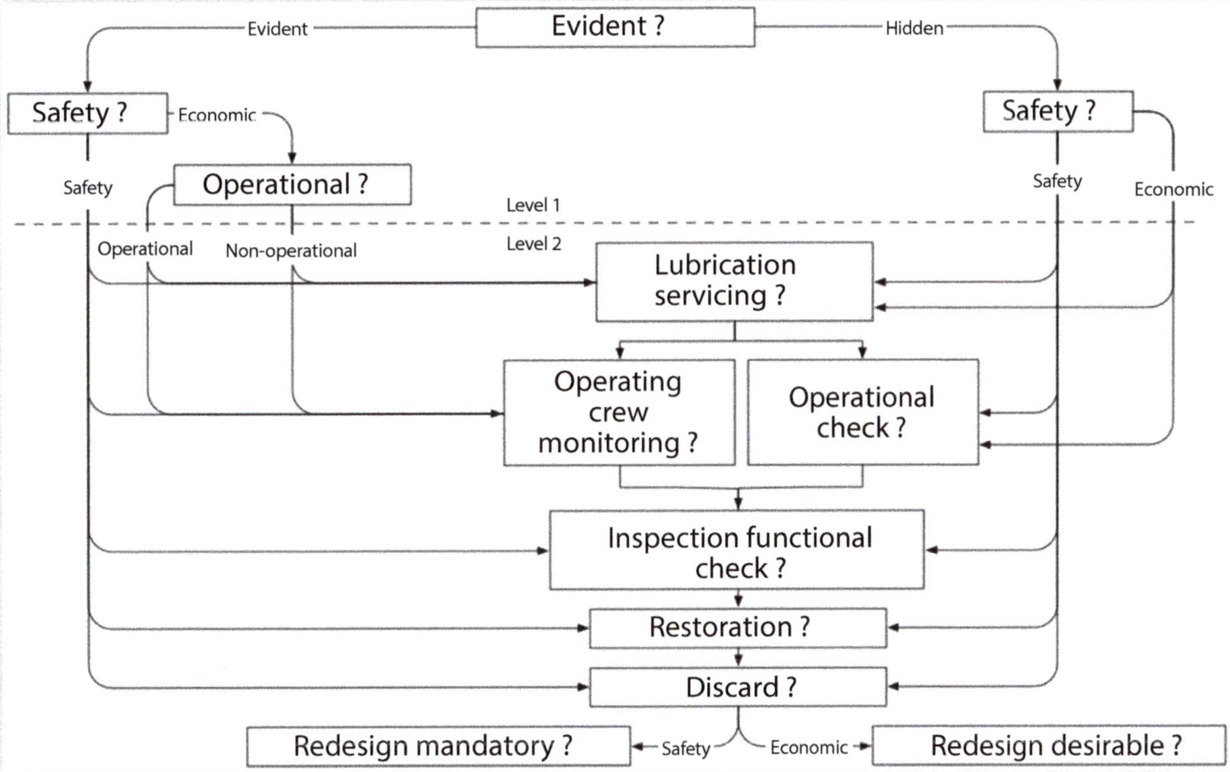

probability of failure to required levels will be harder compared with a failure mode that has operational or economic impact or consequences only. If a failure has safety consequences, then it follows that these have automatic operational and economic consequences as well.

The middle column in the decision logic below level two is where the various types of maintenance tasks are considered. Underlying questions will be applied about the practicality, cost and effectiveness of each of the task types to reduce the probabilities to acceptable levels. If no tasks are relevant, then no scheduled maintenance would be the result where machinery may be allowed to run to failure. However, if the failure mode has safety consequences, other redesign options may be mandatory, which involves modification of the physical design or changes to the operating boundaries, procedures, or rules.

For any aircraft maintenance regime, the MSG-3 is used as guidance: the manufacturers and operators put together an initial new aircraft maintenance regime and present this to the FAA Maintenance Review Board (in the USA) for review and acceptance. There are similar regulators in the rest of the world for the aerospace industry.

PdM emerged as digital control, digital instrumentation, and information and communications technology (ICT) matured. The primary driving force for development has not been the rigorous processes for defining maintenance in MSG-3, but in finding a way of system and propulsion manufacturers to adopt servitization business models where safety is a given, but the operational and economic risks may be better addressed. PdM is a means of accessing data to determine onset of failure that may disrupt operations and drive servitization costs up.

It is only latterly that efforts have been started to explore whether PdM can be exploited more fully to gain "Maintenance Credits," where the reduction of risk of safety-related failures to acceptable levels can be justified with measurable certainty. This work with groups such as the regulators (EASA, CAA and the FAA), operators and manufacturers are working out how PdM may be further exploited.

Key Take-Away Points

- Maintenance is evolving at an increasing rate
- The demands made on maintenance are increasing, with safety, efficiency, environment, ever-improved reliability, and reductions in through-life cost
- There are six patterns of failure: as the complexity of machinery increases, so functional failures tend to be random
- PdM has been adopted in many industries addressing operational and economic factors. Aerospace is attempting to formalize PdM is guaranteeing risk reduction in Safety & Health failures under a revised MDG-3 guidance
- There are many examples in different industries of best PdM practice, where the techniques are adapted to suit different operating contexts and environments
- Maintenance is not just a cost centre, it can significantly add value and extra revenue
- Machines and maintenance are becoming ever more complex: it is necessary to take a systems approach to their analysis and management to ensure success

References

1. Wikipedia Contributors, "ISO 55000," 2020, https://en.wikipedia.org/wiki/ISO_55000, accessed May 2020.
2. BBC History, "A History of the World," 2019, http://www.bbc.co.uk/ahistoryoftheworld/objects/zzVuNAIYSTi4qEApK0o5NQ, accessed June 2019.
3. Nowlan, F.S. and Heap, H.F., "Reliability-Centered Maintenance," Report Number AD-A066579. United States Department of Defense, 1978. Archived from the original (PDF) on 1 August 2013.
4. Moubray, J., *Reliability Centred Maintenance*, 2nd ed. (Oxford: Butterworth Heinemann, 1999).
5. MSG-3, "Operator/Manufacturer Scheduled Maintenance Development (Vol. 1-Fixed Wing Aircraft and Vol. 2-Rotorcraft)". Revision 2018.1, Airlines for America, 2018.

CHAPTER 3

What Is Predictive Maintenance (PdM) and How Does It Fit into a Maintenance Regime?

This chapter introduces PdM, explains why and how it has emerged, and how PdM fits with any maintenance regime. Initially, the context for PdM will be set within a more generalized maintenance system. Not all may agree with the exact context or the definitions used here. The intent is to deliver a working context so that PdM can be better understood. The definitions should be general enough that, when other texts and definitions are presented, they may be recognized and mentally mapped to those presented here. A Maintenance Regime is how an asset-rich organization designs and operates its asset maintenance program. The maintenance regime often includes logistics, spare part management and, where maintenance is carried out (on the line, at a maintenance bay or depot, or at a factory level).

The fundamental contextual setting of PdM is its relation to the whole Maintenance Regime. In this book, the definition of the Purpose of Maintenance is:

> *To preserve functions stakeholders required of machinery – within their defined operating contexts.*

This definition is highly influenced by John Moubray's definition of maintenance [1].

Preserving asset functions has a wider scope beyond preventing and recovering from an asset's physical failure. This is too narrow a perspective to take when considering maintenance and PdM in modern times. The stakeholders of machinery who are the owners, operators, maintainers, users, and ultimately the general public have far wider expectations of assets than in previous ages, and these expectations are constantly growing.

A significant minority of asset-functional requirements are not associated with safety, physical integrity, or performance. There are also requirements for aesthetics, intrinsic value, and the need to enhance the experience of users. For example, enhanced passenger experience is a major differentiator to an airline operating company, so why shouldn't maintenance (such as keeping engine rotors balanced more often, leading to lower levels of cabin noise) directly contribute to this functional requirement? If an airline operating company made a much stronger link between how maintenance could enhance customer experience, then would ensuring in-flight entertainment systems were working properly take on greater importance? How many times have people travelled and been frustrated where the sound does not work, or the jack-plug connectors for earphones do don't either? The point is wider society and the general public also influence the requirements for maintenance. There is great sensitivity to pollution (noise, particulate and greenhouse gasses) that can be reduced by new design and maintenance, where monitoring and maintenance of optimal machinery performance also delivers reduced emissions.

Another significant area where PdM can be directly employed beyond physical failures is maintaining the functional requirements for optimizing performance, efficiency, and emissions. Efficiency is most often thought of as improving fuel efficiency. An example is where aero engines are periodically washed to improve efficiency. Dirt accumulates in the compressor stages which reduces aerodynamic efficiency increasing fuel input for the same power out. Washing restores performance, but also has a trade off because washing may slightly increase corrosion. The PdM system can derive efficiency through several sensor measurements, and then trigger washing when the engine efficiency drops to a predetermined level. The monitoring can also determine how effective the maintenance has been in restoring the efficiency by measuring its increase.

It is also important to break down the words in the definition to enable a full understanding. The next sections will explain what is meant by these terms:

- Functions
- Operating Performance Levels
- Operating Context
- Operating Environment
- Asset Stakeholders

What Are Functions?

Functions are requirements that asset stakeholders expect their assets to fulfil. The functions must include required levels of performance and tolerances where appropriate, which then enables an understanding of what Functional Failure means. Functions and functional requirements should be wide-ranging, going beyond the mere physical operability of machinery. They may include performance and efficiency, aesthetics, look and appearance, residual value of the asset, and the experience of users (crew and passengers in aircraft), etc. The residual value of aircraft is dependent on keeping logs and operating ranges. This defines the extent of what is normal expected behavior. Functions are records of maintenance history, certificates of conformance and an accurate configuration of fitted parts. Without these records, safety cannot be assured, and the economic value of the aircraft is nil. The maintenance regime's functional requirement in this aspect is to maintain accurate and complete records.

What Are Operating Performance Levels?

Functions should include constraints and operating performance settings that define normal often categorized as primary and secondary to help focus the analysis, where a secondary function is one that supports any primary functions.

The Major Influences of Failure

As part of the context for maintenance it is worth keeping in mind what the major drivers of physical failures are when designing the maintenance regime. The predominance of the failure modes and their frequency will be highly influenced by how assets are operated and maintained in their operational environments, as well as design. These factors should guide us to what data is included in the operating context and environment descriptions that need to be integral to a Failure Modes and Effects Analysis (FMEA) (Figure 3.1).

An FMEA is a way of populating a table of information that defines and captures how an asset may functionally fail in a declared operating context and environment, what failure modes and mechanisms are at play, and what the effects and likelihood of the failure are. The combination of likelihood, impact, and detectability of failure may be evaluated in a FMEA Risk Priority Number (RPN) or if a variant of FMEA is used, called Failure Modes Effects Criticality Analysis (FMECA) where criticality is determined using a slightly different process. Criticality is the interplay between likelihood, impact and detectability where the more frequent and higher impact have greater criticality. The evaluation of impact and likelihood are the same factors that are considered in risk analysis processes.

The FMEA has a drawback because it only considers failure modes in isolation from each other. When conducting a full maintenance FMEA study, it is often necessary to consider what happens with multiple failures or, in an industrial plant, what may happen with a failure if other parts of the plant are withdrawn from service. This is an example where systems thinking, and holism apply. It is not enough that all failure modes are

FIGURE 3.1 The major influences on failure.

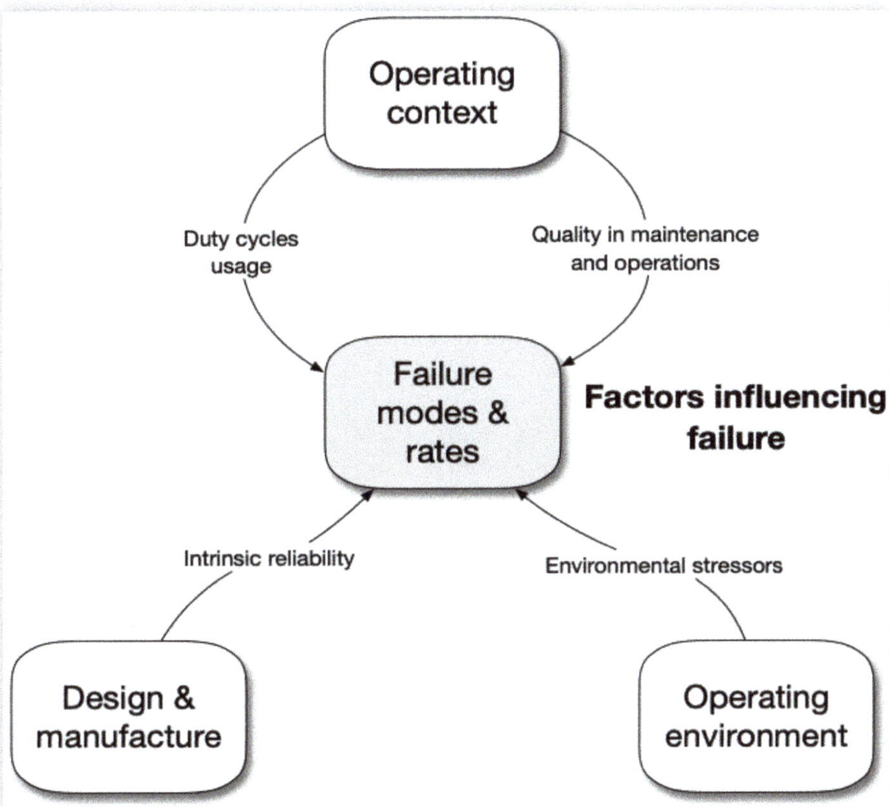

considered in isolation but what may happen when several failure modes may happen simultaneously.

Other tools for considering multiple failures are using Hazops, Fault Trees, and Reliability Block diagrams. These tools are explained in the glossary. In the nuclear industry in the USA, a live Probabilistic Risk Analysis (PRA) based on Fault Trees is necessary to keep aligned with the availability of plant components to show the consequential risks against unacceptable outcomes, such as fission-product release outside of containment to the general public.

There is a big difference between a FMEA conducted during design and one for a maintenance regime. The Design FMEA focuses on the maximum and minimum performance, and is especially geared to failures with safety consequences. The maintenance FMEA focuses on what is required of the machine in its operating context, so the required performance may be less than the design performance. Although safety is paramount and is definitely considered, the maintenance FMEA for PdM also considers failures with significant operational and economic impacts in greater depth. Although a design FMEA also takes into account operational and economic factors, they often cannot anticipate all the corner cases of how machinery is configured and used, and what variations of operating environment may exist.

It is worth considering what the manufacturers recommended maintenance is in the industrial domain. The recommendations are a trade-off between the worst of the most probable machinery use cases the manufacturer knows, keeping through life cost competitive (with similar competitor's machinery), their own manufacturing costs and achieving acceptable competitive levels of reliability. The Operator may use this as a baseline, but then design a far more appropriate maintenance regime by using the RCM or MSG-3 process.

In aerospace, the airline has to get its maintenance approved and the regime is far more rigorous in being approved to ensure safety. However, where systems are not safety implicated but may be vital to influence customer experience (such as in-flight entertainment) then there is more flexibility to determine the maintenance.

Other industries also have strict safety standards, such as the nuclear power and pharmaceuticals where strict approval regimes are enforced for maintenance.

When RCM is conducted retrospectively on industrial plant, it is often found that the design performance of the fitted machinery is not enough to deliver the current functional requirement, especially after other parts of the plant have been modified, or production requirements have been changed. The FMEA process can often prompt design changes and modifications. If a plant or operations go through a staff reduction exercise, it is vital that the MSG-3 or RCM study is reviewed, especially to re-evaluate whether failures are hidden or not.

Writing FMEAs is conceptually simple; the basic rules can be learned in less than an hour. Whether the FMEA is useful or not is another matter. It takes someone with experience and skill (facilitating a team to extract expert knowledge) to write a useful FMEA.

Expert knowledge is also required to understand how deep an RCM analysis needs to be taken in terms of breaking down machinery to constituent parts. Too deep and the analysis will be too expensive and long and the detail at the lowest levels will have little value. Too shallow and important failures may be overlooked that might have safety consequences. The cut-off point is not obvious, but it is vital to get this right. An expert FMEA or RCM facilitator will have the essential tacit knowledge to make the correct decision.

The intrinsic reliability of a product is set by design. Intrinsic design reliability may be reduced by manufacturing, maloperation and maintenance. Maintenance can never exceed an asset's intrinsic reliability, but it has a direct influence on getting close to or achieving it. Intrinsic reliability may only be increased by design change(s) and modification(s).

What Is an Operating Context?

Many people conflate operating context and operating environment together. They are only separated in this text to show some of the distinctions for the sake of clarity.

The Operating Context description should accompany a FMEA or FMECA to provide some of the bounding assumptions that apply to how an asset is operated, and how this may influence failures and behavior. In the experience of the author, an FMEA or FMECA that does not include a description of the operating context/environment has marginal value. If the operating context is missing, then it will be difficult to establish all of the purposes and functions machinery delivers and impossible to fully describe the indirect effects and consequences of functional failure (beyond the machine and its boundary).

The following information is usually required to define an Operating Context:

- What are the boundaries of the asset system under consideration? This may take the form of a diagram. Defining a system and its boundary is a systems-engineering technique.
- References to any engineering drawings or documents should be included. The FMEA or FMECA study will require access to system, ISO or isometric (so the physical positioning of sensors within a system can be seen) and process diagrams, especially in the industrial plant context.
- Identify the inputs, outputs, controlling and influencing factors from the environment acting on the boundary.

  The reasons these are included is to divide machinery and plant up, so work in designing the maintenance regime may be simplified and more efficient. Some parts or groupings of machinery may be more critical than others and may take priority.

  Although it is common practice to divide a whole asset into systems and sub-systems to analyse and define maintenance, the theories of "Systems Thinking" and "Systems Engineering" tell us that other behaviors emerge as systems are connected in a larger asset. This means other properties emerge when systems are joined together and interact. The point is that eventually the design of maintenance must consider the whole asset and the influences it is subjected to from its environment.
- How does the asset deliver business value and how may supply-and-demand fluctuate? The reasons why these are important is that machinery delivers mainly financial and business value. Maintenance must ultimately deliver safe business value. Some industrial plant must deliver in accordance with dispatch commitments (aircraft must meet timetables and fly on time). Maintenance must support these commitments. A high-level description of the impact asset non-availability may have on the organization will enable a wider consideration of what functions machinery delivers and the impact of the loss of those functions in business terms.
- Any high-level asset design assumptions that may be relevant (e.g., the design life and what constitutes the end of economic life of the asset).

  These factors have an influence on how a maintenance regime is designed and run, especially major cadences such as mid-life overhauls and end of economic life.

Is there a desire to have some residual or resale value after the organization decides to sell a used asset?

How does the asset fit in the organization's portfolio of assets and/or plant? What does the asset do, within the context of the wider business? For example, an organization may have a mixed short-and long-fleet. Newer assets may be assigned to more lucrative routes, as an example. Aircraft engines need to work with temperature margins that limit thrust. Engines have to work harder when aircraft operate from airports at higher altitudes. There is a big difference between operating from Mexico City compared with Singapore.

A description of the operating and duty cycles.

Some operating regimes accelerate or increase the influence of factors that impact reliability, performance and usability of machinery. For example, short-haul routes have higher numbers of take-offs and landings per month and consume cyclic life (driving fatigue failure mechanisms) faster than long-haul routes.

- A description of the maintenance cadences (minor and major overhauls):
  - This should include what asset components drive these cadences. For example, some industrial gas turbines have a 4-year overhaul that is driven by restoring the thermal protection coatings on turbine blades and guide vanes. A major overhaul cadence at 8 years replaces the turbine blades and guide vanes with new ones.
  - One way of determining the economic life of a merchant ship is via hull thickness with an assumption it will decrease on exposure to its operating environment (the sea). There are other items that fix the maintenance cadences, so it is worth knowing what these are.
  - Many mining vehicles may have a mid-life engine change, and economic life is defined by the structural health in the vehicle chassis (cracking in the main chassis).
- A brief history of the asset, any modifications or areas of concern in terms of reliability, availability, or maintainability. There are usually significant differences between the as-designed, as-built, and as-maintained states of the asset, which need to be thoroughly understood and documented.
- Any legislation or industry standards that define or influence how the asset may be maintained or operated.

  The conformance to standards and legislation form functional requirements that affect operations and maintenance. Different countries may have different legislation and require compliance to differing standards.
- A description of the organizational breakdown with outline responsibilities for operating and sustaining the asset. This is important to determine who has to make operational logistic, financial and maintenance decisions. Identifying who has the authority to decide and act is vital if the wider organizational system is to work. One example where this would be important is where manufacturers are adopting servitization strategies, where they are seeking service revenues from the after markets based on their products. They may rent or lease their products or guarantee availability and provide their own maintenance for ongoing fees. The adoption of PdM is a very potent way of mitigating servitization risks, and a description of how an organisation has to evolve to embrace servitization and PdM is a vital source for planning and change management.
- A description of the operating environment, especially highlighting which factors will influence the risk of failure and rate of deterioration. For example, exposure to salt and moisture may increase the risk of corrosion. These environmental factors

may be referred to as "environmental stressors." Daily and seasonal variations may also be mentioned. The altitude and general levels of pollution at the main operating airports are important. Mexico City is at 7,382 ft (2,250 m) above sea level which means lower air density which impacts turbine gas temperature (TGT) margins. According to the World Health Organization (WHO), India has six cities cited as the worst polluted, which may increase the rates of sulphidation corrosion [3].

- A brief overview of the current maintenance regime with references to details. In aerospace the main maintenance regime will be explicitly defined, but if PdM is being used outside the official maintenance then it is worth including. Many organizations in other industries may not apply planned maintenance and allow machinery to run to failure.
- The operating context may include the results of a Level of Repair Analysis (LORA). This will include who, when, and where maintenance is done at various levels of depth of maintenance. What maintenance can be done on the line, what maintenance can be done in a hangar or maintenance bay, what maintenance can be done at a specialized Maintenance Repair and Overhaul facility, and what maintenance may be done at a manufacturer's factory? The military has designations of where maintenance is conducted, either On-asset, Line, Depot or Factory.
- In conjunction with the LORA, consideration needs to be given to stores' availability and the logistics system. One of the most important pieces of data needed when applying PdM is the practical and cost-effective lead time in ordering spare parts and comparing this to the P-F interval explained in Figure 3.4. Unavailability of spares is a major reason for planned maintenance disruption, and lead and delivery times must be factored in.

What Is an Operating Environment?

The Operating Environment is the environment that the asset operates in. For aircraft flying at high altitude, the operating environment may be relatively consistent, but differences exist at lower altitudes and airports where air pollution may vary, and corrosion rates may be accelerated. Other similar environmental stressing factors can include altitude (of an airport), proximity to the sea with salt-laden moisture, or proximity to desert conditions with abrasive sand. Offshore oil and gas platforms have a harsh environment to operate in with design features to suppress the effects of corrosion.

A useful reminder of the different influences of Operating Context and environments is to consider the case of the Tornado multi-role combat aircraft in Europe. The Royal Air Force operated Tornados as high-altitude interceptors in Northern Europe and as low-level strike aircraft during Operation Desert Storm in Iraq. The maintenance demands varied widely on the same asset between these two scenarios.

A further illustration of how failure is influenced is to consider aircraft assets operating in either the civil or military sectors. Military aircraft used on the front line are generally designed optimised for performance, whereas civil aircraft are generally designed optimised for fuel efficiency and long-term reliability in mind. This does not suggest reliability is not important in military assets, but to ensure superior military capability against adversaries' design trade-offs are different. These design differences influence the failure modes and the rate of failure for differently designed assets.

Who Are Asset Stakeholders?

The functional requirements required of assets are derived from the asset stakeholders. Some of the stakeholders are direct with explicit requirements, other indirect with implicit requirements. Stakeholders may include:

- Asset owners, leasers, financiers etc.
- Asset Operators (including those who may have financially invested)
- The asset operating and maintenance crew
- The asset users (passengers in the case of civil aircraft)
- Industry regulators and standards groups
- The media at large
- The general public

A Taxonomy of Maintenance Tasks

Having discussed influences on failure the book moves on to discuss another aspect of maintenance that provides context for PdM. This is to describe a taxonomy of maintenance type tasks to describe where PdM fits.

There are a variety of maintenance task types listed in Figure 3.2 below. Many of these task types are also included in the MSG-3 decision logic in Figure 2.5. This taxonomy provides a context to show where PdM fits within the wider maintenance domain.

FIGURE 3.2 Taxonomy of types of maintenance task.

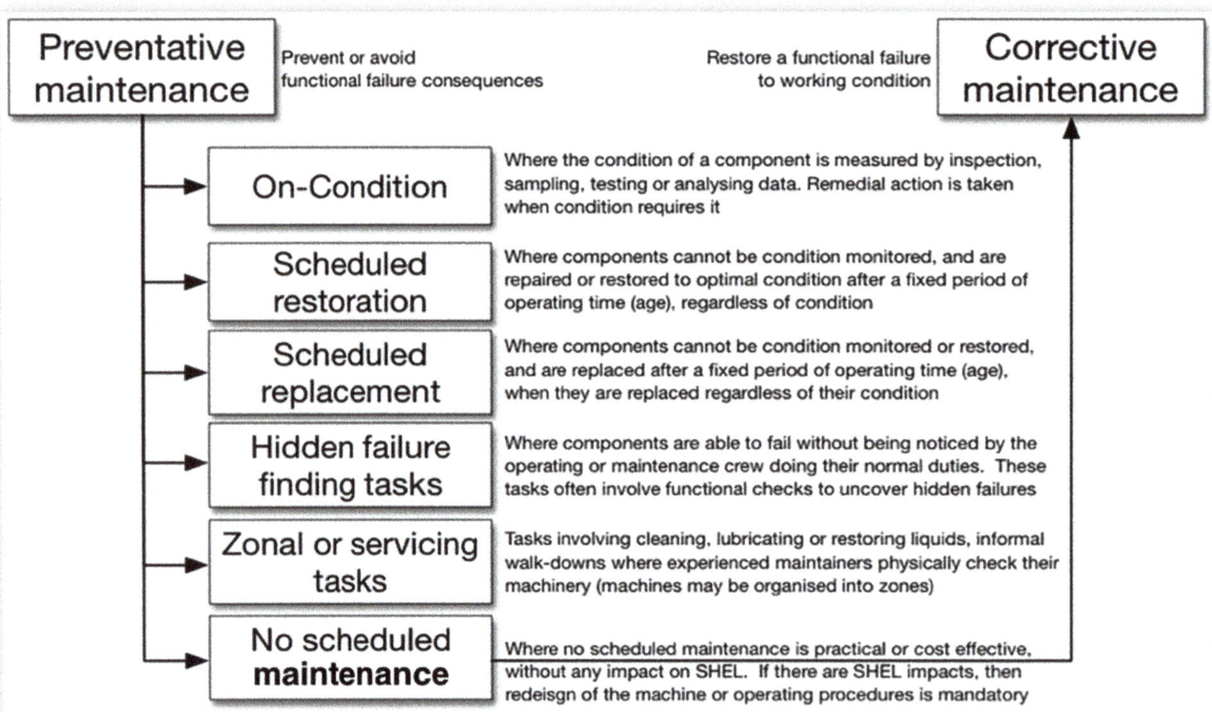

PdM can never replace the entirety of other types of tasks within Preventive Maintenance, because there are physical constraints that limit PdM. These limits will be explained below, but as PdM capabilities improve due to the emergence of new technology it may take over a higher overall percentage of all maintenance tasks. It is also important to know what type of maintenance task is applicable to which failure pattern. This knowledge enables the selection of appropriate maintenance task types as some have limited applicability. This principle is embedded at the heart of the RCM philosophy.

Although RCM recognizes 6 distinct patterns of failure (see Figure 2.4), these six patterns may be abstracted into three main groups: age related (wear-out), random failure, and infant mortality.

Age-related failure is where the probability of failure increases with age. Random failure is where the probability of failure is nearly constant despite the age. The component is equally likely to fail no matter what the age is. Infant mortality is where the probability of failure is higher when the age is lower. It is caused by quality problems: the appropriate maintenance response is to conduct Root Cause Analysis (RCA) to determine causes that can be rectified and to focus on those rectifications. The rectification actions may take considerable time, such as waiting for equipment modification. If so, it may be sensible to apply on-condition maintenance (including PdM) as a palliative action to try and contain the effects of infant mortality.

The Taxonomy of Maintenance

Maintenance may be decomposed into various types of activities or tasks as follows. Within the explanation a description of which failure patterns apply will be mentioned:

Figure 3.2 has a reference beside "No scheduled maintenance," to SHEL this stands for Safety, Health, Environmental and Legislative compliance. This is explained in detail later on in Chapter 3.

There are many versions of the maintenance taxonomy that vary from this figure in other textbooks. PdM is a type of on-condition maintenance. Some may distinguish PdM as being pro-active and not part of preventative maintenance. These other definitions may be motivated by marketing to differentiate the capability, but any preventative maintenance task is proactive. There have been some discussions about proactiveness and maintenance in focusing on defect elimination. This relies on having a failure history and conducting root cause analysis (RCA) to enable action to eliminate some causes of failure to improve achievable reliability. This activity is valid but not covered in this book. This book will explain each type of maintenance, so that readers can reconcile other taxonomies if and when they encounter them.

The detailed explanation of maintenance type tasks shown in Figure 3.2 are as follows:

Preventative Maintenance

There may be some disagreement in marketing circles that PdM and so-called Proactive Maintenance (associated with defect elimination and redesign) are separate from Preventative Maintenance. This *Aerospace Predictive Maintenance: Fundamental Concepts* posits that these are marketing attempts to emphasize differentiation of these techniques to increase sales. They are not truly new branches split from Preventative Maintenance.

On-condition maintenance is where a component's condition may be measured, and a prognosis made of how much time is left before a component deteriorates so much that it cannot fulfil its functions. This calculation determines the Remaining Useful Life (RUL) of the component. PdM is a subset of on-condition maintenance.

Component condition is monitored by fixed sensors continuously with software doing the diagnostic and prognostic analysis. By continuously we mean the fixed sensors do not need intervention by people to acquire the data independent of the sampling rate. PdM may sample a subset of data from fixed sensors at rates different from what it is sensed. This could mean data is sampled at a rate between milliseconds to once every week, or at discrete times within an aircraft flight or sector. Traditionally, on-condition maintenance has been conducted by periodic inspections or surveys that determine condition and initiate corrective action as necessary. On-condition maintenance is applicable to any of the six basic failure patterns. On-condition maintenance enables the prediction of functional failure, but not the onset of failure in the incipient failure model (Figure 3.4). It does not strictly prevent failure but allows us to avoid functional failure and its consequences.

Scheduled Restoration is where a component is refurbished, repaired, or cleaned to restore its condition or performance to as close to new as possible at a time when the probability of functional failure increases or performance drops to an unacceptable level. The task is completed at a set interval, regardless of condition, that may be based on elapsed calendar time, operating hours, cycles, or some other accumulating count associated with operations. The "regardless of condition" implies that the condition cannot be economically or effectively measured, and on-condition maintenance is, therefore, not optimal. Scheduled Restoration is only applicable to components failing with the age pattern.

Scheduled Replacement, often termed Scheduled Discard, is where a component is discarded and replaced with a new one, regardless of the condition of the original component at a time when the probability of failure increases to an unacceptable level. As common with Scheduled Restoration, the "regardless of the component condition" implies on-condition is not economically viable or practical. Scheduled Replacement is only applicable to components failing with the age pattern (see Figure 2.4).

Many maintenance regimes that have not been undergone the RCM process make the mistake of applying Scheduled Restoration or Replacement tasks to components exhibiting random failure. This is not effective at reducing unreliability, it merely increases through-life cost.

Hidden Failure Finding is where a component may exist in the failed state unnoticed by the operating crew or maintainers undertaking their normal duties. This situation is perilous where the failed component has a protection or standby function. If the device has failed and a second event requires the protection or standby function, then the consequences of this multiple failure is severe. In most modern systems, safety or protection systems are designed to be "fail safe," in a condition that is instantly recognizable by the operator. Hidden failure-finding tasks may include scheduled physical or functional checks or inspection. An example of a safety device failing in a hidden manner may be a safety relief valve fitted to a pressure vessel that has failed in the shut state (its normal operating state). The relief valve would fail to function on a subsequent overpressure excursion in the pressure vessel, which may cause the vessel to catastrophically fracture. Hidden failure-finding tasks may consist of physical or functional tests or on-condition inspections.

In the FMEA discussion above, it was pointed out that the FMEA does not really consider multiple failures acting together that may cause higher levels of unacceptable consequences. However, the FMEA should consider hidden failures a consequence of two failure events as an exception. The other tools such as Hazops, Fault Trees, and Reliability Block diagrams should also be used to account for multiple failures.

Servicing and Zonal Checks: These types of checks may include lubrication and cleaning tasks, as well as experienced maintainers conducting more informal walk-downs and machinery rounds. The five human senses combined with the brains of the experienced operator and maintainer are among the best systems for detecting and recognizing abnormal behavior and early signs of problems. These checks may be related to the rate of consumption of consumables. Operator or maintainer walk-downs are a type of on-condition task.

No Scheduled Maintenance: This may be a deliberate choice where maintenance is not possible or not cost-effective. If this is the case and unacceptable residual risk of failure exists, then redesign will be mandatory if it is safety implicated or desirable if operational or economically implicated, otherwise failure risks and criticality are small, and items may be left to run to failure where corrective maintenance will then be applied.

Although the next sections are not included in Figure 3.2, the author considers it is worth including here as many other books include this calling it "proactive maintenance." This activity is not a maintenance task, but an approach and effort expended by the operators and maintenance crew to improve reliability. There is a framework called Total Productive Maintenance (TPM) that includes these activities. Predictive maintenance is a subset of on-condition maintenance that exploits digital technology.

Redesign and Defect Elimination: Ongoing efforts to recognize where there are benefits in either improving intrinsic reliability (by redesign and modification) or actively removing causes of failure of defects.

Predictive Maintenance (PdM) is an advanced form of condition monitoring and on-condition maintenance that exploits digital technologies in sensing, communication, and data processing.

PdM may be highly automated using acquired sensor time series data to diagnose when assets or components begin to fail from a failure mode, determine the condition of an asset, and then to prognose remaining useful life (RUL) in which any remedial action needs to take place.

PdM may also be different from traditional on-condition maintenance because:

1. It enables continuous monitoring of assets through its fixed sensors, which does not require any scheduled task to sample or gather condition-related data, although it may merge data from these other CBM-type tasks to get a richer picture of asset condition.
2. The timeliness of data processing should be optimized to align with the ability to act and the impact of functional failure. PdM offers the opportunity to do near-real time processing if this is justified.
3. The PdM system can provide alerts to different people who manage assets. Alerts and advice may be output to local line maintainers if remedial action is urgently required. Where longer P-F intervals are involved, the PdM system may inform planners and managers who can pre-dispose resources and plan for recovery, liaising with the operational managers when it is least disruptive to withdraw assets from service. PdM systems that feed information to operators should be treated with caution for fear of overloading them with information (mixed with alarms and warnings) if assets degrade to abnormal states. This is described in more detail further down in this chapter.
4. PdM may combine data from separate sources to gain a richer picture of condition. For example, combining vibration data with oil-debris data increases certainty that bearing incipient failure is diagnosed. Much of this combination may be done in an experienced maintenance manager's head but having a system that does this automatically allows the experienced maintainers to focus effort on improvement, as well as helping knowledge transfer.

Deeper Explanation of PdM

It is necessary to consider two main models for PdM, best explained by two diagrams. The first model is called Long-term Degradation, which covers time spans lasting months to years. The second model is called Incipient Failure Detection where the P-F interval may last milliseconds to months in duration. The two figures look very similar at first glance, but the practical differences are profound. With Long-term Degradation there is no failure inception point: the asset begins to degrade as soon as it enters service or as soon as it is exposed to its operating environment (for example, a ship is launched before it is fitted out and the hull is exposed to corrosion in a seawater environment).

It should be noted that the curved cumulative decay shown in both diagrams may not represent many failure modes and how they degrade. These types of curve have been used in most textbooks describing on-condition maintenance and are repeated here for familiarity. Other degradation modes (such as damage accumulation or fatigue degradation) may be more linear. A minority of the traces or loci of degradation may even show improvement: for example, if cracking in a structure (such as a gas turbine fan blade) initiates, it may change the resonance of that structure which may be seen as an improvement depending on where vibration sensors are sited.

It is noteworthy that a similar list in an existing standard CBM in JA 1011 showing the pre-requisites (described above in this section) for PdM (the guidance document for RCM) does not explicitly mention that variance in the times to failure needs to be reasonable. The variance is important as it increases the practical difficulties in prognosis and increases uncertainty that remaining useful life is sufficient to take remedial action.

It is also noteworthy that in long-term deterioration it may not be possible to measure the deterioration and condition until late in the life of the component. An example may be low-cycle fatigue that causes cracking in structures and may only become apparent by traditional NDT or inspections, when the cracks grow big enough and appear on the surface of the structural material, which could be relatively late in the degradation cycle. This may be best illustrated by using two examples:

- Long-term degradation in pipe wall thickness. It is possible to measure pipe wall thickness at any time in the degradation cycle, and so the condition can be quantified for the whole degradation, by scheduled on-condition checks.
- Long-term degradation in structural integrity. It is possible to determine the emergence of structural material cracking caused by fatigue by NDE inspections when cracks appear near and at the surface of the material. This occurs late in the overall degradation cycle, where possible 80% or 90% of the useful life has already been consumed. The NDE inspection periodicity needs to be adjusted to a smaller P-F interval. Because degradation might not be detectable, critical structures may well have an associated damage-accumulation model built, where PdM continuously monitors stress cycles and applies formulas that predict the condition to augment the NDE regime. Such a model increases the confidence in estimating the degradation of condition (Figure 3.3).

The second mode to be considered is when a failure may initiate at any time during the in-service life, and there is a relatively smaller time delay until the condition degrades to a point where functions are lost. At first glance this chart looks very similar to Long-term Degradation, but the major differences are that the Incipient Failure Detection has a point of failure inception (perhaps some kind of shock event that initiates failure). The second major difference is the time scales over which the majority of the failures occur:

FIGURE 3.3 Long-Term Degradation.

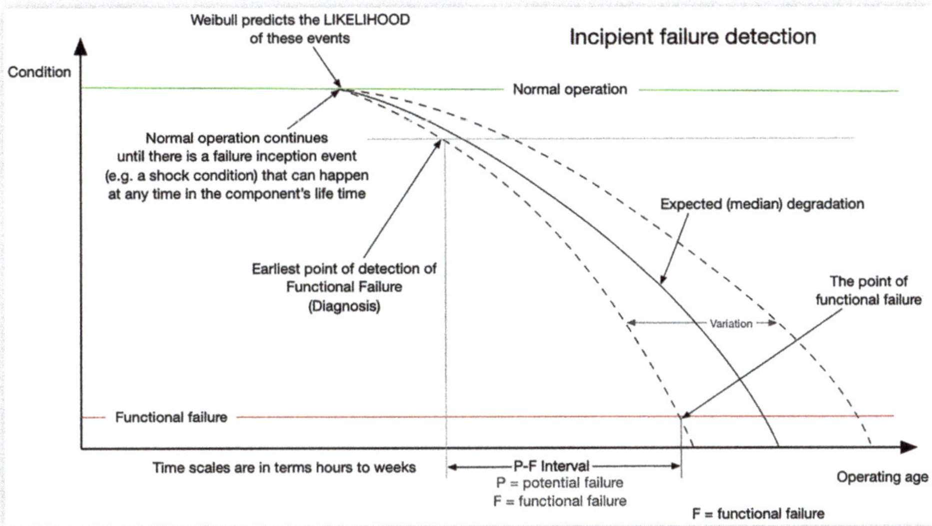

the vast majority of Incipient Failure Detection times are much less than the Long-term Degradation. This is illustrated in the following figure:

FIGURE 3.4 Incipient Failure Detection.

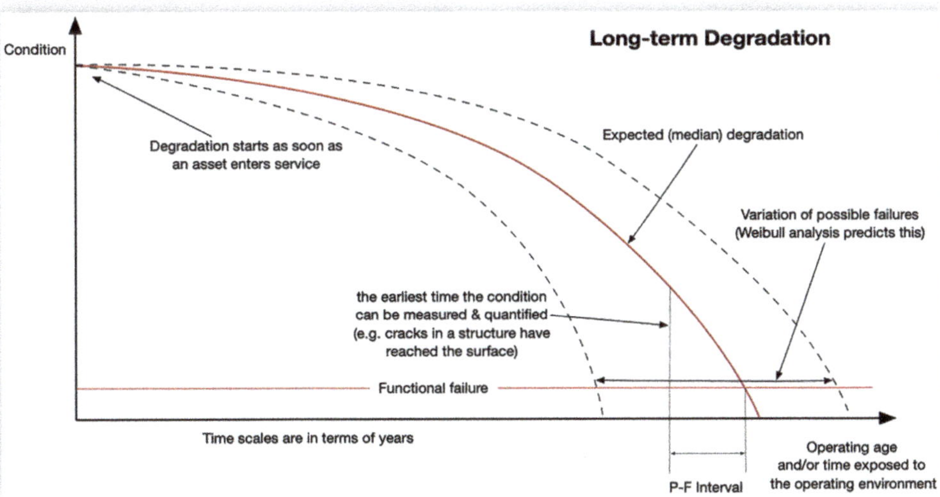

The periodicity of the sampling rate, if PdM continuous monitoring is not available, needs to be a fraction of the P-F interval. With PdM that has continuous monitoring of fixed sensor data this monitoring frequency is not an issue.

Successful PdM in common with condition monitoring and condition-based maintenance relies on certain physical characteristics of how assets or their components fail.

There are a number of pre-requisites that must be met if any condition monitoring and PdM are going to be successfully applied. These are:

1. There must be a time period between inception of failure and the eventual functional failure.
2. The failure (or potential failure) must be observable during the failure lifecycle. It is sometimes feasible that the inception of failure itself may be observable or is known.
3. The time period between potential failure and functional failure (often called the P-F interval) must be:
    a. Consistent with acceptable variance between other similar failure events of the same failure mode, on the same type of components. The variance needs to be reasonably small so that the estimate of remaining useful life may be accurate.
    b. Long enough to allow for planned recovery and the predisposal of resources, to execute the most effective and efficient recovery preferably timed allowing the minimum disruption to operations.
4. If condition is not continuously measured, then the periodicity of condition sampling must be a fraction of the expected P-F Interval.
5. The effort and cost to conduct PdM must be less than the avoided cost of the consequences of unplanned functional failure. Simply expressed the Return on Investment (ROI) must be positive. If the likelihood of failure is small, then the cost of applying PdM may be large before detection, implying the effects or consequences of failure in these cases are large. In any organization seeking to adopt PdM, the ROI must be greater than 3, with a rapid breakeven within a year or two. Often in the author's experience, ROI up to 10 is often achievable.

The difference between the P-F interval (illustrated in Figure 3.4) and remaining useful life (RUL) is that the P-F interval is the time between the earliest possible time a failure may be diagnosed to the time of functional failure. The RUL is the time from any point within the failure lifecycle after diagnosis until the functional failure. After diagnosis, the RUL reduces as time passes toward the functional failure point.

Levels of Diagnostic Capability

There are a range of PdM systems varying in maturity and capability. The simplest PdM includes a straightforward means of anomaly detection and has no failure mode diagnostics or prognostics. When judging the maturity and effectiveness of the PdM system, a set of questions to determine the extent of the analytics should be adopted based on the discussion below.

The simplest type of anomaly detection is based on single-sensor readings with a set threshold so that, if the sensor reading exceeds it, the system alerts the user. The sensor data trends tend to be uncompensated for environmental or other influencing parameters also captured by available sensor readings. For example, if the PdM system were monitoring an air-cooled cooling system, then analysts should compensate for daily and seasonal ambient temperatures. Often the system sets PdM alarms on important parameters below that already set by installed warning and alarm systems. The PdM system is in competition with the alarm systems and usually is not particularly effective. An uncompensated, single-parameter-threshold detection system for detection of anomalies is a recipe for excess false positives and is not worthy of being called a PdM system. If false positive alarms are regularly received then the motivation is to set the threshold higher, which then risks missing genuine abnormal excursions.

Most immature PdM systems only use threshold exceedance to flag an anomaly. This may be too late and not give enough time to act. Anomalies should be able to be classified from threshold exceedance, step changes (that do not breach exceedance values), and gradual rates of change. If a residual signal has a steady increase or decrease, then anomalous behavior can be alerted long before a threshold value is exceeded.

Mature anomaly detection has techniques beyond simple threshold exceedance.

One of the most effective PdM diagnostic strategies is to use a "model of normality." The simplest model of normality is the same sensor reading from a similar machine in the same operating environment. It is important that the same operating environment is shared to do this. On an aircraft with lots of redundancy this is easy: for example, engine-to-engine comparisons are an effective strategy. Both engines on a single aircraft share the same operating environment. The advantage of using equipment-to-equipment comparisons is that they generally automatically compensate for environmental and influencing parameters which considerably reduces variance. This has a beneficial effect in reducing false-positive rates. Using a model of normality, the predicted reading of the dependent variable can be subtracted from the observed reading of the dependent variable, resulting in a residual that can be trended over time or other variables. The trend of residuals can be regarded as a "deviation from normal behavior" signal, making it very simple to observe anomalies. The concept is illustrated in Figure 3.5.

Figure 3.5 shows an example of both using equipment-to-equipment (E-to-E) comparisons or where a model is used. If using the E-to-E, the other equipment is substituted as the model of normality in the Figure 3.5. It is important to ensure trend data is labelled properly, with identifiers that include the asset or machine, plus the "tag" of the sensor(s) supplying the data. In an industrial setting, the author has often discovered discrepancies between data-tag identifiers in a data-historian database and sensor identifiers in a process or ISO engineering drawing. Sometimes different identifiers are

FIGURE 3.5 Model of normality with residual variable.

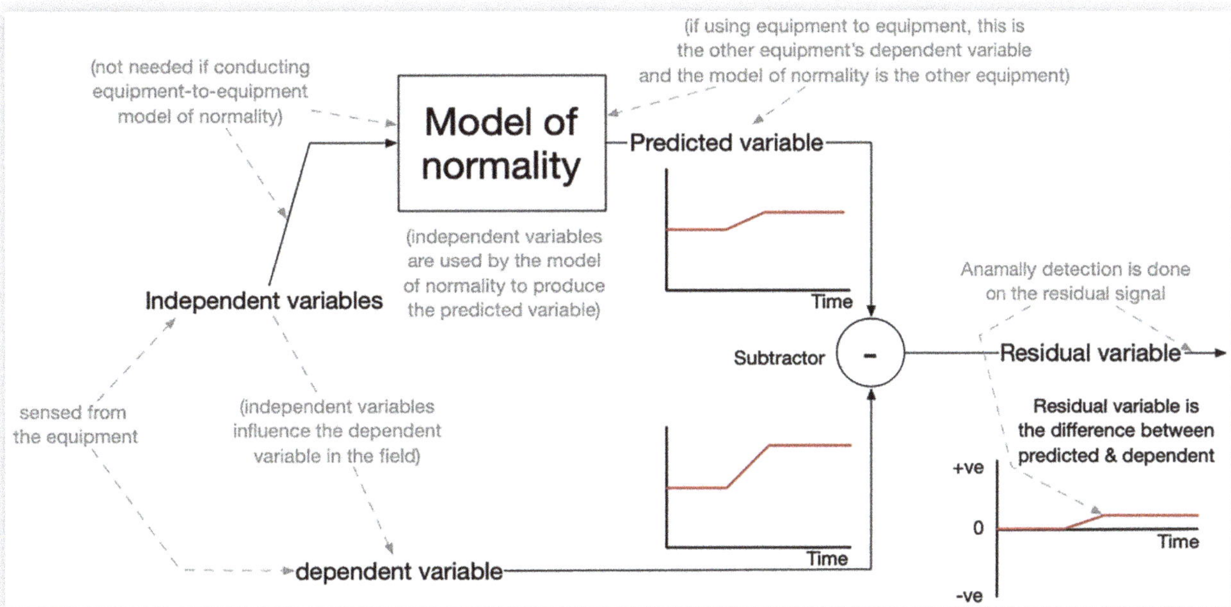

used in the databases compared with the drawings. Mistakes in identification can lead to false positives and lost credibility in the PdM system.

The model of normality may be physics-based, or could be a ML or AI model, where the influences of the independent variables on the dependent variable is learned from historical sensor data. Physical modelling may be preferable over data-driven models for auditability and trust by the engineering domain team because they may be reflected on and understood from engineering first principles. However, the ease and simplicity of building data-driven models, especially in Python, has considerable advantages in rapidly deploying and using models. In a mature PdM system, training and test data preparation, model building, testing, deployment and automated triggering for re-training is usually an automated pipeline of execution. There are shortages of PdM experts and others such as data scientists, as well as experts in the customer organisation. Automation is key to success.

Years ago, only manufacturers had the intellectual capacity to build accurate physical models, and they had a monopoly in using models of normality. Today, anyone with access to complete time-series data can build-data driven models: the modelling capability has been effectively commoditized by Python etc. (see Chapter 10). It must be stressed that data-derived models are only representations of the data used to train them. The data used for model training needs to include operations at every part of the normally operating envelope, and needs to include other environmental influencing data to ensure accurate representative modelling. For example, in an industrial context, the full range of seasonal and daily variation in ambient temperature may need to be included. An aero engine will require altitude and air temperature. The differentiating factor today for conducting PdM is not modelling, it is access to representative sets of machinery data.

Having said this, the author does not endorse PdM modelling from a purely data-driven approach. Engineering domain knowledge is still a vital part of validating and verifying that models are working as required. Some companies might claim they can fully automate processes of normality modelling, novelty and anomaly detection, and diagnosis. This is a false claim, because they often rely on the tacit engineering knowledge within their customers to diagnose and validate the outputs. This dependence on expert interpretation is not normally mentioned in a sales pitch, and any organization with limited engineering domain expertise may be overwhelmed with managing the volume of anomalies that are reported. Automated diagnosis is required to deliver a scalable solution. There is also overhype of data science being a silver bullet solution that can deal with low levels of data quality. People building or adopting PdM need to be sceptical and alert to extravagant claims.

It's an obvious realization that a digital twin can be used as a sophisticated model of normality for many dependent variables, as well as a model for prognostics.

Diagnostics

Anomaly detection has some value, but a mature PdM system needs to be able to differentiate specific failure modes for identified components. Diagnostics is a classification problem.

A definition of terminology is required to understand the build-up to diagnosis:

- A novelty is a detected deviation from known normal behavior. The novelty is not yet confirmed to be an anomaly and may be a case of operating within a previously unknown area of normal behavior that is not recognised by a model of normality. A novelty might be caused by a benign event such as performance improvement after restorative maintenance.

- An anomaly is a detected deviation from normal behavior that is known to be abnormal. It is not known whether the anomaly is a symptom of a failure mode. Some immature PdM systems are limited to only detect anomalies, which may be diverting the PdM user into insignificant alerts.
- A symptom is an anomaly that is a known constituent indication associated with a diagnosis of a particular failure mode. A symptom is significant and needs to be brought to the attention of a PdM user.
- A set of known symptoms correlates to the diagnosis of a failure mode. Not all symptoms may be present at the same time. Some symptoms may only present themselves later as the failure degrades through time. The time delays may be quantified (as a distribution) and treated as symptoms in their own right. A confirmed diagnosis and a prognosis need to be alerted to the PdM user, and the machinery operating and maintenance authorities to start planning for recovery.

The shift from anomaly detection to diagnostics may be achieved by understanding the way a set of anomalies from different sensors change through a timeline in relation to other anomalies. A set of anomalies behaving in a certain repeatable pattern that are associated with a failure mode can be thought of and termed as symptoms of failure. Taking an engineering perspective of symptomatic anomalies, it is possible to make expert predictions about the likely physical effects of the failure mode, and this knowledge can be used to predict what other physical effects may be presented in other sensor data. If sensor data has been retained that covers older occurrences of the same failure mode, then the data can be mined to confirm or refute these predictions. Part of the PdM system requirement is to be able to mine the historical event (failure or maintenance events) and sensor data.

Many failure modes may share similar symptoms so, to isolate the failure modes, it is necessary to discriminate between the symptoms. The nature of how symptoms present themselves and their timing in context with other symptoms may be used as discriminatory factors to help isolate the failure modes. For example:

> *A filter may be diagnosed as blocked when the flow through the filter is reduced, combined with an increase in the pressure drop (the delta-pressure or DP) across the filter increases. Both of these symptoms MUST be present to* unambiguously *diagnose a blocked filter. In many blocked filter cases, only one of these parameters is available to a diagnostician, but if this is the case false positives may occur. An example may be if a system has a blockage downstream of a filter causing a reduction in flow readings through the filter, which would not be caused by a blocked filter.*

Some symptoms may or may not present. A specification for a diagnosis may include a set of symptoms that includes the probability that symptoms will appear, including any delayed timescales. Each symptom may also include a vector representing the magnitude in the trend's change and direction (plus or minus) of the symptom's difference from expected normal behaviour will be for each failure mode.

In the real world there is never a set of perfect sensors to get unequivocal symptoms to diagnose failure modes unambiguously. The engineering domain experts must look for other clues and compensate with other sensors that infer information relevant to a more accurate diagnosis. For example:

> *In the case of the blocked filter, if filter DP is the only reading available, it may be combined with other indications such as system pump parameters or other temperature readings (that vary if the flow of fluid is not normal) that may help the diagnostic discrimination. The diagnostician should try and infer the fluid flow through the filter in this example.*

There are certain techniques that can detect the onset of failure earlier than others for certain failure modes. For example, for a failing rolling element bearing, oil, oil debris, and a vibration analysis may be the earliest and most sensitive symptoms. These should be followed by increased bearing temperature and, just before functional failure, audible noise coupled with temperatures so high a maintainer cannot keep their hand on the bearing housing. Differing techniques may be applied if it is necessary to gain sufficient P-F intervals to plan recovery, and minimize disruption.

Having more than one technique for diagnostics also facilitates prognosis, as the symptoms appear in a predictable time between each other.

Mature diagnostic capability is characterized by:

- Multivariate sensor (parameters or features) normalized to account for normal behavior and machine state in the trend, to detect novelties. In data science, each sensor time series data trend that has correlation to detecting a failure is called a "feature."
- Detecting different behaviors in features such as threshold exceedance, rates of change, step changes (a step change is a high rate of change). Is the feature trending high or low? What is the percentage of change from normal (using the residual signals)? Note that step changes may be observed in parameters after a short outage period because of the effects of carrying out planned maintenance. The PdM system needs to classify and recognize the effects of planned maintenance and ensure they are not reported as diagnosis of a failure mode.
- Recognizing a novelty and classifying it either as an anomaly or not. An anomaly may be thought of as a candidate symptom of a diagnosis.
- Differentiating the timing and sequence of observed anomalies. When do some anomalies present themselves in a timeline? Some symptoms may be delayed. How can this information be used to discriminate? How can this behavior be used to help prognostics? The time is a feature in this analysis. (The timeline itself is a feature). The time to presentation of anomalies may also be a signature in its own right and may also be useful in prognostics.
- Analyzing a set of multiple anomalies as a symptom. A set of anomalies that may have direction and magnitude may be present over time and can be grouped together as a diagnosis. Such a defined set of anomalies to a diagnosis may then be labelled symptoms. Some symptoms will always be present when a particular failure mode initiates, some others may or may not be present but may appear later, or the magnitudes may vary over the P-F interval period. Other symptoms may only present themselves later on in the failure, but if the delay for their appearance has a reasonably small variance, then the time delay before they appear is useful information in both diagnosis and prognostics.
- Providing a level of confidence or accuracy of diagnosis. The diagnosis should show a measure of certainty that the diagnosis is correct. This can be based on the strength and number of symptoms present combined with historical statistics of diagnostic accuracy.

It should be possible to build up a knowledge base of the symptoms of diagnostic. An example may look like:

Failure Mode: Blocked Filter; Machine: XXX; Equipment: Filter Identifier

- Normal operating range DP xxx to xxx
- Symptom: Slow rise (ramp) in DP Normal expected rate = xxx per xxx from renewal
- Symptom: DP exceeds threshold xxx (Normally expected 3 to 5 months after renewal)
- Symptom: Lube Oil pump (xxx) discharge pressure residual slow rise in line with Filter DP rise. Note: The residual implies that a model of normality is being used to model the lube-oil pump.
- Symptom: Both pump bearings (xxx and xxx) slow rise in smoothed residual temperature in line with filter DP rise. The temperature change will be less than one degree F, and both bearing temperatures will change together. (Note: it may be possible to use each bearing as a model of normality for the other bearing, smoothing the residuals also implies that the change in temperature is small and would be difficult to detect in noisy data).

This specification could be expressed as a set of rules in a decision tree and encoded in the PdM system to detect the failure mode.

The following figure shows elements of novelty, anomaly, and diagnostics in a mature PdM system (Figure 3.6):

FIGURE 3.6 Functional blocks associated with a mature PdM system.

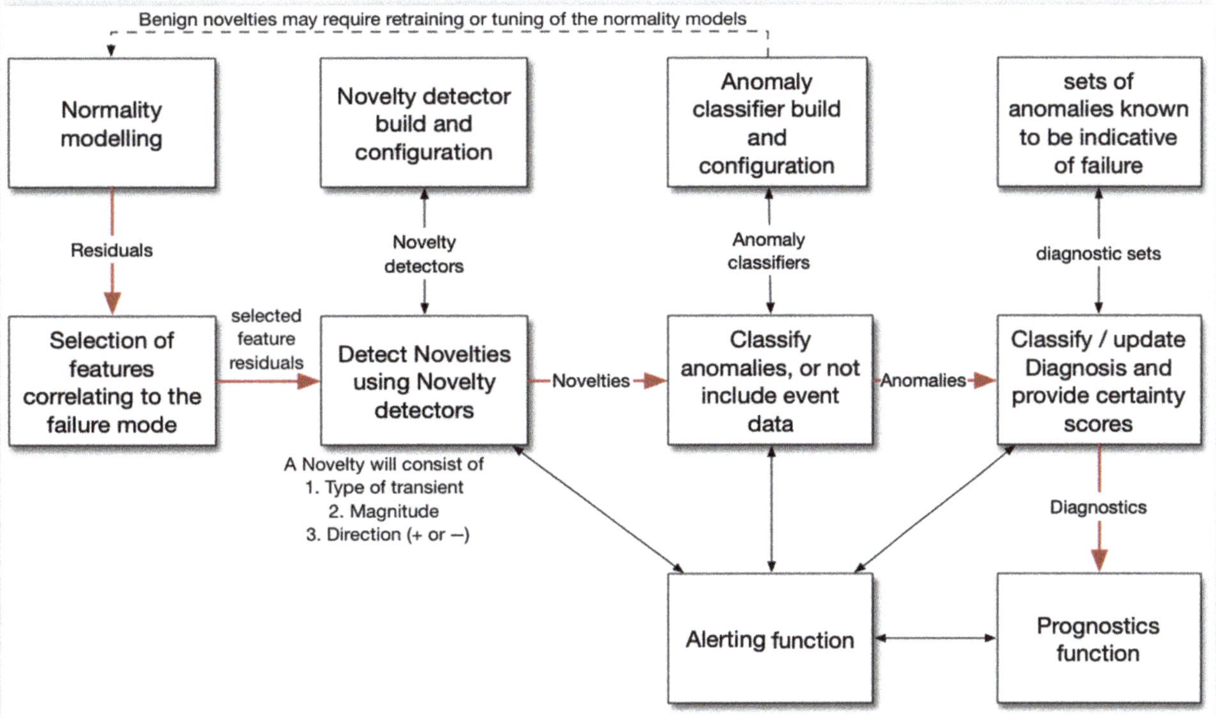

Diagnostic Effectiveness

Measuring diagnostic effectiveness is mostly achieved by a confusion matrix and Receiver Operator Characteristic (ROC).

The confusion matrix records whether the true state of a system is classified correctly over a number of failure events. The classifier's performance can be recorded as below (Figure 3.7):

FIGURE 3.7 The confusion matrix.

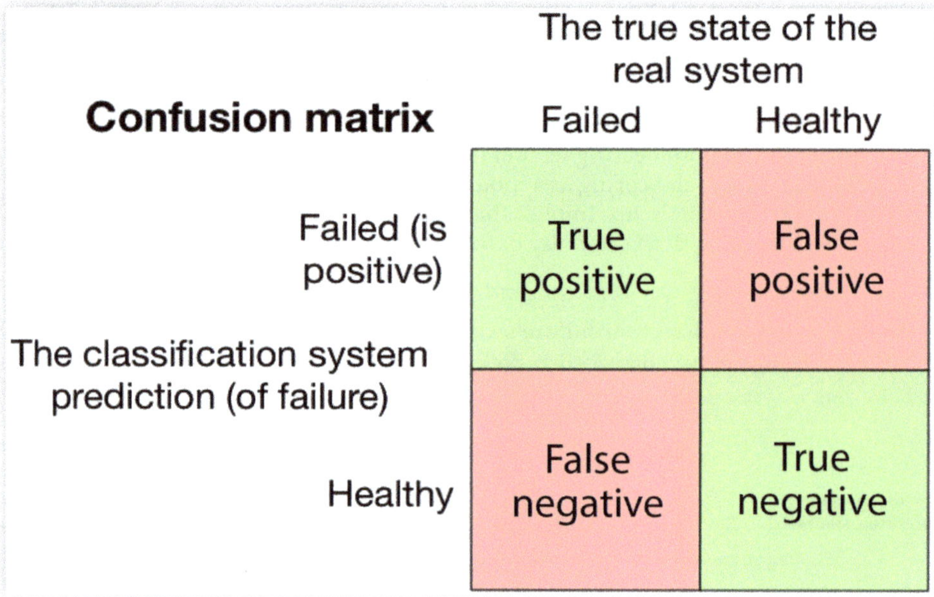

The terminology of the confusion matrix may be fuzzy because of the words used in the goal of the classifier. If the goal of the classifier is to detect system failure, then this is the "positive" in the matrix and the healthy state of the system the classifier detects is the "negative." The words seem contrary; the trick is to always remember the goal of the classifier. In the confusion matrix figure, the green squares are good and the red ones are bad.

Each quadrant in the matrix represents the following

- A **true positive** is when the classifier correctly detects the system's failed state
- A **false positive** is when the classifier indicates the failed state when the real system is healthy
- A **false negative** is when the classifier indicates the healthy state when the real system has failed
- A **true negative** is when the classifier correctly detects the system's healthy state

The false positive rate (FPR) is the number of false positives divided by the number of negative events.

The true positive rate (TPR) is the number of true positives divided by the number of positive events (Figure 3.8).

FIGURE 3.8 Receiver-operator curve (ROC) characteristic.

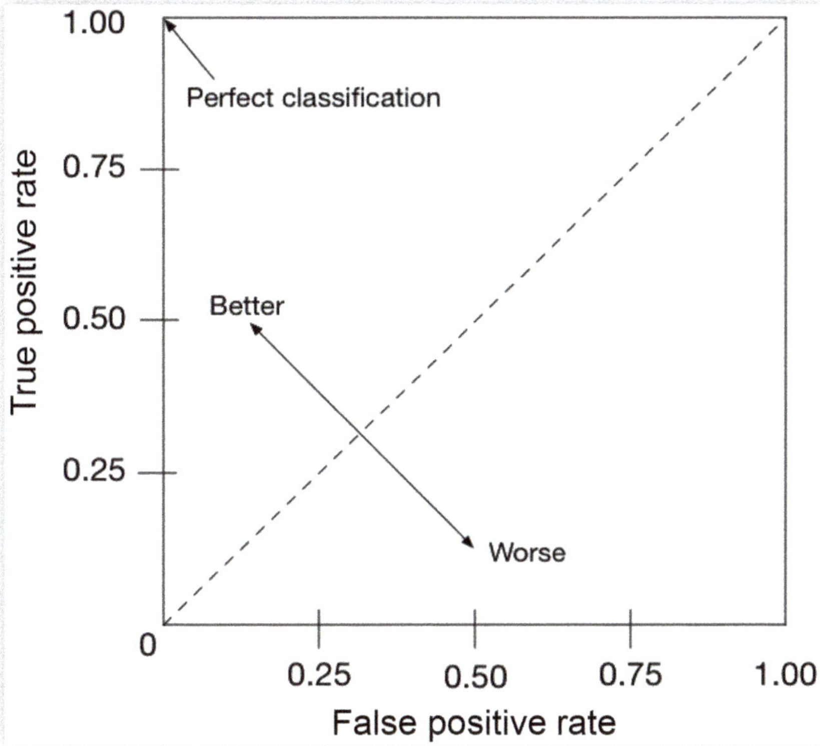

The following text is adapted from the Wikipedia explanation of ROC:

An ROC is a chart with an x axis measuring FPR and y axis the TPR as x and y axes. This shows the trade-offs between true positive (benefits) and false positive (costs). Each prediction result or instance of a confusion matrix represents one point in the ROC space.

A hypothetical situation may be where a failure mode has safety consequences, where the classifier missing a failure event (a false negative) is unacceptable because the safety consequences are catastrophic. In this case, it may be acceptable to have a small number of false positives as an optimum trade-off because the wasted effort in remedial work is better than missing the safety implicated failure. On the other hand, the optimisation and tuning of the classifier may be reversed in a situation where the operational disruption may be more expensive than the direct consequences of the failure itself (which are only economic). In this case, a trade off to minimize false positives, accepting a small number of false negatives, may be preferable. Diagnostic specificity or sensitivity may be tuned and optimized.*

Prognostics

Much seminal work on Prognostics has been done by the NASA Ames Research Centre.

Prognostics is a regression-type problem and is distinct from Diagnostics (Diagnostics is a classification problem). Prognostics is used to estimate the remaining

* Specificity and sensitivity are defined in the glossary at the end of the book

useful life before functional failure as well as describing the effects of the functional failure, and what recovery may entail. It is helpful to use an analogy of medical diagnostics and prognostics to understand the difference and what the content should include. A doctor may tell you:

> *"You have measles (The Diagnosis): you will be off work for approximately a fortnight, by when you should have fully recovered. You will remain contagious for a week, so it is not advisable to mix with other people until after this time (The Prognosis)."*

A PdM system that does not include prognostics is an incomplete system. The types of model identified in the diagram to the right of the boxes marked "Type" are explained below (Figure 3.9).

Type 1 Models

Weibull model: The Weibull distribution is a continuous distribution most often used in reliability engineering to fit failures in a population of the same components. Other distributions can also be used occasionally. The distribution may be represented by three parameters: Location, Shape, and Scale. However, in most cases, Location is not used, as most components do not degrade before they are used in-service. If a sample of failure data is captured for a component and the age of the data at the time of failure is known, then there are estimation algorithms that can be used to fit the data to the Weibull distribution. This can then be used to determine the likelihood of failure for components at a specific age using a derived cumulative distribution function graph. Some

FIGURE 3.9 Types of prognostic model, their data and what they address.

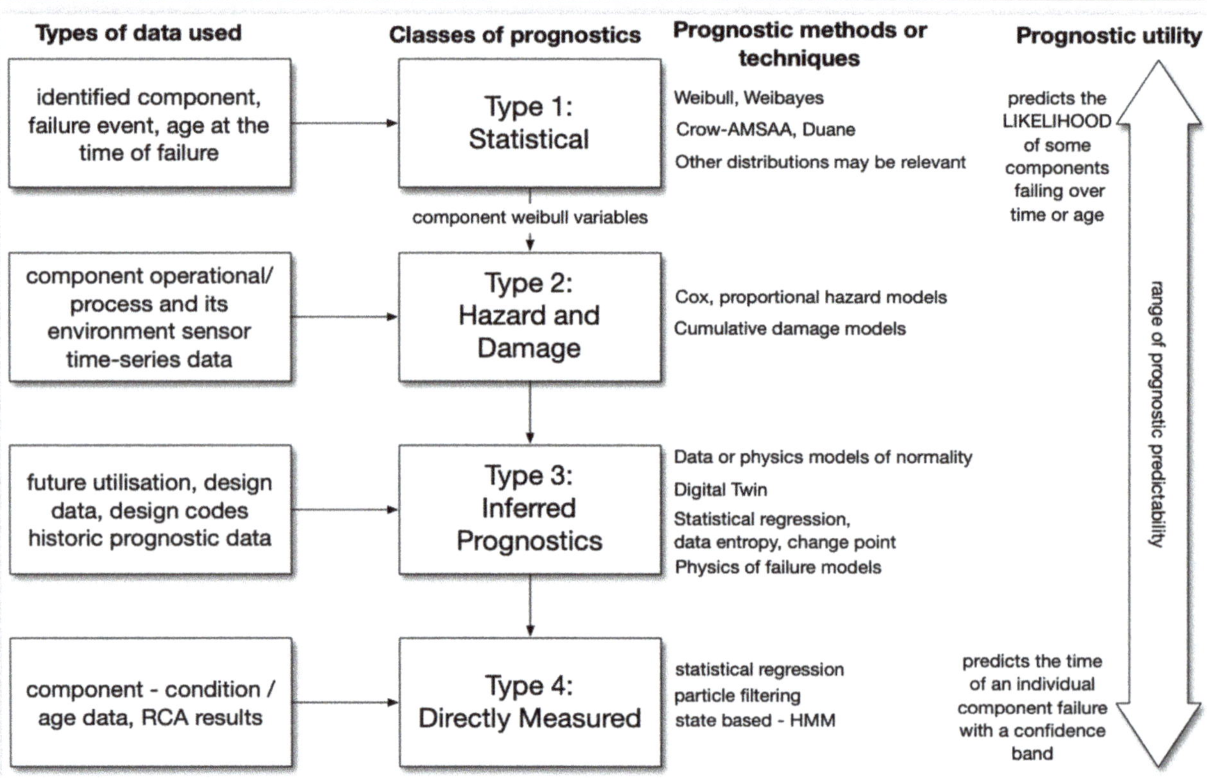

components degrade from the time of manufacture (e.g., rubber hoses and their time since cure date) may require the Weibull location parameter to be used, or age since cure date may be used as the start of service date for a 2-parameter Weibull.

Weibull may be used to prognose long-term deterioration that defines component life but is in practice rarely used to determine incipient failure deteriorations. There is no technical barrier to using Weibull analysis to model the distribution of time-of-failure inception (or diagnosis) until functional failure, especially for PdM systems that use continuous monitoring. If historical events are used, then some compensation for the condition of the removed parts may need to be employed, because at the time of change the components may not be in the functionally failed state.

Weibull analysis is valuable for another reason inherent in the RCM process, which is the selection of maintenance-type task, dependent on the basic patterns of failure that parts fall into. This will be discussed more in Chapter 5, where a deeper understanding of how RCM and PdM interact will be given. In mature systems with many assets of the same type there may be sufficient failure data to conduct Weibull analysis. This can determine whether the failure patterns are infant mortality, random or age related through the Weibull shape parameter used in the RCM process to select which maintenance tasks are applicable. The Weibull results can also be used for prognostics (for long term deterioration) and may be used in RAM (Reliability–Availability–Maintainability) simulations to inform supply chain and logistics decisions to ensure supply of spares as well as supporting through-life cost simulations that can support capital expenditure projects for upgrading or modifying assets.

The Duane [4] or Crow-AMSAA [5] are similar techniques and were originally developed to show reliability growth during testing phases of a new product. They can, however, also be used during the in-service phase of a product lifecycle where failure events, dates, and ages can be sorted in date order, and plotted on a graph with logarithmic axes. The data may show change of slope that indicates whether reliability is getting better or worse. This can act as a warning that operational or maintenance quality is changing with worsening failure rates.

Type 1 prognostics can be used to predict the likelihood of what proportions of the population of components are likely to have failed at a certain age. It is possible to use Crow-AMSAA to predict likely time of failure for single components, but accuracy of the prediction depends on the number of previous failure events.

When the author worked in the MoD, in the UK submarine support team, we used Cumulative models such as Crow-AMSAA in preference to Weibull. This was because we were supporting a flotilla of four boats, that were new to service and the number of failures per equipment were small. The uncertainty in Weibull analysis was too large to base reliability improvement decisions on. Crow-AMSAA provided a earliest warning that reliability trends were of concern and needed early intervention.

Type 2 Models

Proportional-Hazard or Cox Models [6], are able to build on the Weibull distributions by adding other influencing continuous or discrete data that correlate to age at failure. Other continuous data may be environmental stressors (moisture, temperature, etc.) or other cyclic data that may influence deterioration.

Damage accumulation models may take cyclic data (such as engine starts or shutdowns, or electrical air circuit-breaker operations) that define the age of component failure. Thermal cycling may cause low-cycle fatigue where the temperature transients may cause differing expansion and contraction rates in structures made of different materials,. The more rapid the change in temperature, the more stress is imparted and the higher the damage accumulated.

PdM may be used to feed type 2 models with continuous sensor data for environmental and dependent parameters.

Type 3 Models

This is where models are used to infer prognostics either using physical models or those constructed using data, both of which may also be termed digital twins. Some models may simulate damage accumulation itself; for example, modelling structural crack propagation. Type 3 and 4 prognostics is where PdM plays a major role.

Another useful method used in both diagnostics and prognostics is to use change-point analysis and information entropy in the time series data. Change point and information entropy are related concepts used to determine when underlying things have changed (such as an initiation of failure and its local physical effects picked up by sensors). Information entropy is distinct from thermodynamic entropy and was developed by Claud Shannon as part of what is called Information Theory [9]. Information entropy informs us how much information is in an event (which might be the next time a data point is recorded in a time-series trend). If the event is deterministically predictable, then the information content is reduced, if not the information is increased. Entropy is the measure of the uncertainty and information in these events—entropy may peak if a failure initiates where another sensor reading may deviate from what is expected. These techniques are part of the arsenal of algorithms and techniques used by PdM to detect anomalous behavior and help establish reference points of change that are useful in prognostics.

In diagnosis, a change point may show where failure is initiated, or a symptom or novelty presented itself. In prognostics, change points can be used with a stochastic approach as part of an inferred stochastic approach. A stochastic approach occurs when a set of parameters that vary (or the set of parameters change their state) randomly over set periods of time, may be used to determine whether known states (an example being defined by the diagnostic rules above) is reached. The entropy or change state knowledge may be mixed in with the diagnostic rules and may be used to help identify states of degradation for prognosis. It has already been described that diagnostic symptoms and their time of presentation also provide useful data for a stochastic model for prognostics.

Type 4 Models

Direct observation or measurement of the direct mechanisms of failure may be used as a basis to predict RUL.

Other models can use a stochastic approach where there are more obvious state changes with temporal relationships between the states. This may also work where some of the diagnostic symptoms of failure only show after a time is spent deteriorating. For example, in the case of a bearing failure, oil debris, oil and vibration analysis may pick up the earliest symptoms (enough to conduct a diagnosis) but, as deterioration proceeds, audible noise and/or local heat after predictable periods of time are other later-developing prognostic symptoms (changes of state) that can contribute to increasing the certainty of diagnosis and prognostic regression estimation.

Both Type 3 and Type 4 models can take either inferred or directly read prognostic data and conduct a statistical regression on them. An often-used prognostic technique is Particle Filtering, which is a combination of Monte-Carlo Simulation and Kalman Filtering to estimate the regression of a failure.

Fallacies and Hype Surrounding PdM

PdM is an emerging technology. In common with other emerging technologies, recognized by Gartner with its famous Hype Cycle, (Ch 2, Reference 7), there is a lot of marketing hype and misplaced expectations. This section aims to discuss some of the claims and clarify how they might or might not represent true value.

The 'Real Time' Label of Superior PdM Misunderstanding

An often-claimed advantage of PdM is that it happens in real time, delivering results to remote (from the assets) PdM users, within seconds of data being sensed from distant and mobile assets.

A question here would be, "what do you mean by real time?" Real time may have a requirement in a protection system to invoke the safety system within milliseconds of detecting the failure, or where a failure that requires remedial action by an operator within an hour can be declared within 5–10 s of detecting the condition. If a PdM system is sending data from a remote and mobile asset (such as an aircraft) to a central point where analysis is conducted and remedial actions decided, then the system may be constrained by communications and network latency, which one would never be able to predict without having installed a real-world system and understood its operation.

While the ability to monitor running equipment in real time is an advantage, and detection of plant transients or instantaneous defects are necessary for plant operators (aircraft pilots, are a good example) to both operate normally and deal with abnormal conditions, a wider perspective should be taken. It is really necessary to understand the system boundaries that separate other condition-monitoring systems that feed protection, alarm, warning, and condition-based maintenance systems. Figure 3.8 illustrates these sub-system boundaries. In general terms, on-board systems on the left of the figure require real time, the PdM system on the right-hand side of the figure needs to be timely, but most often not real time. The later chapter discussing Integrated Vehicle Health Management also helps answer this question.

Also, note the two discrete regions of the remote (off-board) PdM system, each corresponding to the two models of PdM, the incipient failure and long-term degradation regions illustrated in Figures 3.3 and 3.4.

The fallacy is that the remote PdM system does not benefit from being real time. Any alerts of anomalous behavior or failure diagnosis must, however, be timely. The costs of achieving real time (milliseconds responsiveness to a remote site), especially where mobile assets are concerned with the effects of internet latency, are likely to be prohibitively high. If real time is going to mean anything useful, it needs to be quantified and qualified. A reason for near-real time (e.g., receiving an alert within a minute or two) is when it needs to be received while an aircraft is in-flight by a ground-based response system, so it may pre-position maintenance staff and spares to meet the aircraft on arrival at its next landing airport. A minute or two delay in transmission to receiving the alert may be perfectly acceptable. A PdM user needs to ask the question "Is this level of latency classed as real time for their purpose?" (Figure 3.10)

How often will the severity of the alert require immediate investigation on landing, and does this warrant the cost of implementing a real time system? Perhaps aircraft bird strike or an environmental act of God may require it. On the other hand, other failure modes not caused by acts of God that suddenly take a turn for the worse or are initially diagnosed with a severely low condition suggests diagnosis capabilities need to

FIGURE 3.10 Sub-system boundaries where on-condition principles apply (Source Author, a simplified copy is also included in API 691, contributed by the Author).

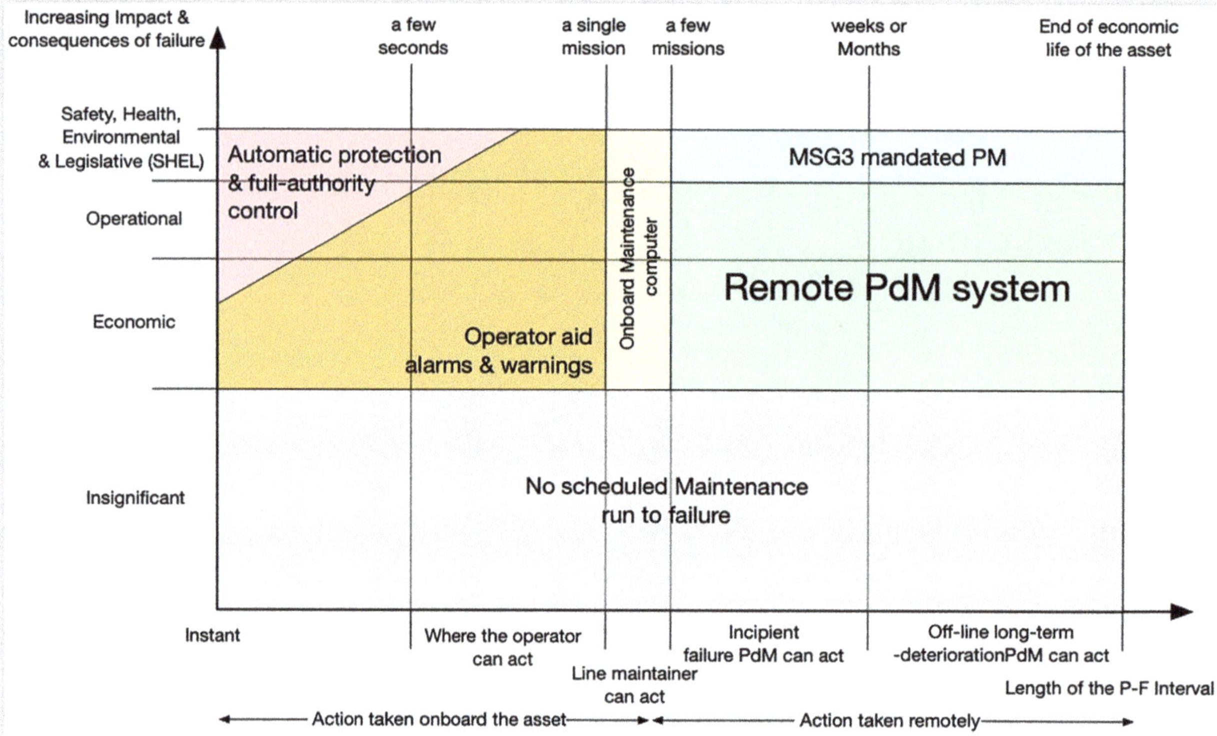

be improved. This implies a continuous process of learning and incremental improvement is needed in operating a PdM system.

A Data-driven Approach Negates the Need for Engineering Domain Knowledge

This claim is only partly true: the PdM system is not in a position to replace domain experts, but may make them more productive and effective. Using the diagram in Figure 3.4 above, it is possible to see a generalized picture of the volume and completeness of data compared with the impact of failure. This book also describes the Resnikoff Conundrum (Chapter 9) which explains that data is rare for safety-implicated failures, because those failures are rare. They are rare because too many safety failures would result in the asset type being withdrawn from use, as legal and societal pressures will not allow unsafe machinery to be used. The other main reason why safety impact failures are rare is because the majority of effort by a product design team have been expended to eliminate or severely reduce the risk of safety-related failures. We as a society would not expect anything less.

This implies that there is more useful data about failures that have ever decreasing impacts of failure. The trap for a purely data-driven approach to PdM with no engineering domain knowledge is that the most satisfying analysis with complete data is possible on failures that do not matter. If the pure data-driven approach is taken, then what matters may not be known. Domain expertise is necessary to establish what matters, and this is usually and usefully contained in a maintenance Failure Modes and Effects Analysis (FMEA).

Note: The text deliberately mentions a Maintenance FMEA. There are many types of FMEA which means some (such as a Design FMEA) are suboptimal to build a maintenance regime.

What Is the Difference between Condition Monitoring and Condition-Based Maintenance?

These terms are often used interchangeably but, while there is a close relationship, monitoring condition is an activity that may be conducted alone without any intervening actions. The condition that warrants maintenance action is a secondary consideration that depends on prognosing remaining useful life (RUL), the operational context, and the impact and consequences of an end failure. Other condition monitoring may re-establish a probabilistic baseline in a risk-based approach to maintenance, where further operation for a defined period is judged to be at acceptable risk levels of functional failure.

It is possible to start monitoring the condition of a machine before it is operated in service but exposed to its operating environment. If the failure mechanism that is intended to be monitored is primarily driven by the operating environment, then monitoring of its condition should begin at the launch. A good example is measuring the thickness of a ship's hull while being fitted out and then when exposed to sea water after its launch, thus measuring its gradual corrosion. While the marine paint protecting the hull is working optimally on launch, it gradually dissipates and allows corrosion to begin. The general point here is that some monitoring needs to start as soon as a (partially built) asset is exposed to its operating environment, which may be before it is operational.

Immature Systems Are Sold as PdM Systems

If a system is capable of only detecting anomalies, it is a very immature PdM system. If we consider a system where engineering experts are able to access all sensor data, then they will be overwhelmed with data and will only be capable of "firefighting" emergent failures. A massive increase in effectiveness is delivered with anomaly detection, but in most organisations the volume of anomalies will still overwhelm the engineering experts. A PdM system limited to anomaly detection is not scalable to gain significant business value.

PdM systems need to include automated diagnosis linking several anomalies to an identified failure mode, of a component(s) at a deep-enough indenture to facilitate effective recovery and effective automated prognostics to enable remaining useful life to be quantified and in which remedial action can be planned and resourced, minimizing the impact of operational disruption.

Many salespeople may not understand what full maturity is and claim what is in effect an anomaly detection system, as a diagnostic system. The buyer of such systems needs to question the vendor to establish the maturity of the system before committing to buying or installing. Chapter 8 provides a high-level specification that can be used as a basis to assess such systems.

Wasteful Number of Inspections

On-condition monitoring including PdM sometimes has to use scheduled inspection and periodic sampling (hand-held vibration surveys, or oil or oil debris analysis). If managers do not understand that the periodicities are a function of the P-F interval, and they observe possibly many hundreds of historical inspection or sampling tasks that have not discovered failures, then they are often tempted to reduce the frequency of inspections to try and save costs. This is a serious mistake especially where the failure being monitored may be infrequent but may also have a high impact. The PdM system needs to be linked to the FMEA data to maintain the technical justification for why the scheduling periodicity has been selected. With the advent of IIOT, it may become practical and cost effective to fit sensors so that the PdM system can conduct automated and continuous monitoring.

How Does PdM Impact Maintenance Planning and Scheduling?

What Is a Maintenance Schedule?

A maintenance schedule is a database often managed by application software termed a Computerized Maintenance Management System (CMMS), showing all of the preventative maintenance where the call-up periodicities are quoted. The schedule should also have gone through maintenance packaging where the call-up frequency is aligned, so nugatory repeating work can be avoided. The maintenance schedule applies to a fleet of assets, managed by the operator.

What Is a Maintenance Plan?

It is thought that one of the best aspects of planned scheduled maintenance is that the future maintenance and material demand is highly predictable because it is either calendar or operating hour-based, which makes maintenance scheduling easy. In reality, many maintenance scheduling plans are disrupted by unexpected failures.

In a Maintenance Regime that is predominantly "on-condition" based, the variance of degradation and the incidence of random failure events means a maintenance plan is more dynamic, less predictable and needs more automation. The key to reducing the impact of this dynamic behavior is having sufficiently long P-F intervals that enable planning and pre-disposition of resources to be made. There may be physical constraints on the length of the P-F interval that frustrate this, but if planning or logistics processes are improved it may be possible to accommodate shorter P-F intervals, which may allow system wide trade-offs.

Another way of mitigating this is to have a 'future looking view' showing

1. When scheduled maintenance is due
2. What is the current and prognosed remaining useful life (RUL)?
3. The probability of failure of components coming from Weibull analysis of failure events
4. The probability of failure of components using damage accumulation models

Digital Twin

One of the recent advances in PdM is the adoption of digital twins (Ch 5, Reference 8). These models may be seeded by data captured from the real asset and the digital twin operated as a virtual copy. The digital twin can be made up of the most sophisticated and detailed design models a manufacturer has available, but the processing cost and time may be prohibitive. If a detailed digital twin is exercised covering the complete set of the asset's normal (and known abnormal) operating states and the data from the modelled sensors is collected, then it is possible to build-data driven models from that output data, and use derived models for the digital twin shadowing requirement. The derived models will be much simpler and have very high computing performance, allowing them to run in true time.

Digital twins may also be stimulated with known errors and exercised in abnormal circumstances to try to gain insights into unknown areas of behavior that result in the specification of new symptoms the PdM system may be set up to detect. It may be possible that the positioning and orientation of sensors may also be optimized using digital twin models.

Key Take-Away Points

- The purpose of maintenance is to preserve functions—not just preserve machines
- A physical asset may be functionally failed, yet not be physically broken.
- Maintenance and PdM are complex systems both involving complex machinery, software and people. A "systems" approach is required.
- PdM is a specialized sub-category of on-condition or Condition-Based Maintenance (CBM).
- PdM does not supersede or replace RCM. RCM promotes the adoption of PdM if the right questions are answered and technologies considered in the RCM decision logic.
- An FMEA (or FMECA) is an essential reference for the design of a maintenance. regime. The FMEA for a maintenance regime is different from a design FMEA.
- PdM should be defined by an RCM process (for industry - MSG for aerospace), along with a maintenance FMEA.
- Real-time PdM systems tend to be hyped: PdM systems must be timely to ensure defective machinery can gracefully withdraw from service, recovery planned, and operational disruption minimized.
- Mature PdM systems include novelty, anomaly detection, and diagnostics and prognostics. If any of these three major sub-systems are missing from a PdM application, then avoid it. Beware the salesperson that does not know the difference between PdM anomaly detection and diagnostics or conflates the two.

References

1. Moubray, J., *Reliability Centred Maintenance*, 2nd ed. (Butterworth Heinemann, 1999).
2. https://en.wikipedia.org/wiki/Systems_engineering, accessed February 2019

3. https://en.m.wikipedia.org/wiki/List_of_most-polluted_cities_by_particulate_matter_concentration, accessed 20 March 2019
4. http://reliawiki.org/index.php/Duane_Model, accessed August 2019
5. http://reliawiki.org/index.php/Crow-AMSAA_(NHPP), accessed August 2019
6. http://reliawiki.com/index.php/Proportional_Hazards_Model, accessed August 2019
7. https://www.gartner.com/smarterwithgartner/5-trends-emerge-in-gartner-hype-cycle-for-emerging-technologies-2018/, accessed August 2019
8. https://en.wikipedia.org/wiki/Digital_twin, accessed June 2018
9. https://en.wikipedia.org/wiki/Entropy_(information_theory), accessed September 2019

CHAPTER 4

How Does PdM Fit with Integrated Vehicle Health Management (IVHM)?

IVHM was invented because it was recognized that Predictive Maintenance (PdM) had developed and been focused on aircraft sub-systems in isolation of each other. The engines, avionics, structure, and other systems had PdM developed independently by their manufacturers. The IVHM idea is that PdM should be a whole, integrated platform approach justified by a business case. An IVHM system should be designed and built using an open and tiered architecture with a systems-engineering approach as a full platform capability, and be a basis for enhancing or replacing traditional maintenance, delivering maintenance credits. The open architecture enables sub-system manufacturers to build PdM systems that can share data with the other platform systems, although the concepts of IVHM have been developed in aerospace the principles are applicable to any industry. In different circumstances there may be a need to stretch the meaning of vehicle (in the IVHM), to include any industrial plant. The word "vehicle" may be mistaken to only apply to mobile assets; this is not true. A deeper explanation about IVHM may be found in [1].

Many industrial plants have some existing PdM systems including periodic vibration analysis using portable vibration-sensing equipment, oil and oil debris analysis with samples sent to laboratories and possibly other non-destructive testing (NDT) or examination (NDE). These results are usually reviewed in isolation of each other and any other PdM system. Often the IT systems used to gather the data and produce the results are standalone and the data is proprietary and cannot be shared between different IT systems. This misses the opportunity to combine the information to gain a richer picture of machinery health and condition following the ideas behind IVHM.

IVHM aligns with other standards such as:

Mimosa [2]: This open standard has been produced by a consortium of industrial organizations, including initial funding from the US Navy. Mimosa incorporates the Open Systems Architecture for Condition Based Maintenance (OSA-CBM) that is defined in ISO-13374 [3] as a functional architecture. The ISO standard is split into four parts. Mimosa defines a logical architecture with UML and SQL schemas for data and interfaces used by the ISO standard and provides a reusable architecture based on SQL data management.

Mimosa also defines the Open Systems Architecture for Enterprise Application Integration (OSA-EAI) standard, which defines data structures for storing and sharing information about equipment. The scope includes equipment configuration, reliability, and maintenance. OSA-CBM uses many elements of OSA-EAI.

Another important project Mimosa runs jointly with the OPC Foundation (the Mimosa and OPC Foundation CCOM OPC Working Group: [4]) is the OPC UA standard for industrial data acquisition. Transfer and storage are having an interface standard developed so that OPC UA may have access to the wider scope of Mimosa asset data. OPC UA is a mature open standard implemented in most data historian applications, and provides a means of collecting and centralizing sensor data from a wide range of industrial data systems where there is a myriad of lower level data networking standards. Data historians have been developed over many years to efficiently acquire, store, and move industrial data. The use of data historians is an area where aerospace lags behind the capabilities of other industries. Data historians were dealing with very large data sets before Big Data emerged (Figure 4.1).

The OSA-CBM stack is a similar abstraction as the SATAAL breakdown described in Chapter 7. The SATAAL acronym stands for the major functional areas PdM decomposes to. It stands for "Sense," "Acquire (data)," "Transfer," "Analyse," "Act," and "Learn."

It is worth being aware of both because SATAAL and OSA-CBM provide valuable but different perspectives on PdM; OSA CBM being more technical and SATAAL more functional. Using them both enables a better PdM system to be architected.

FIGURE 4.1 The OSA-CBM (ISO 13374) functional block diagram (Source Mimosa).

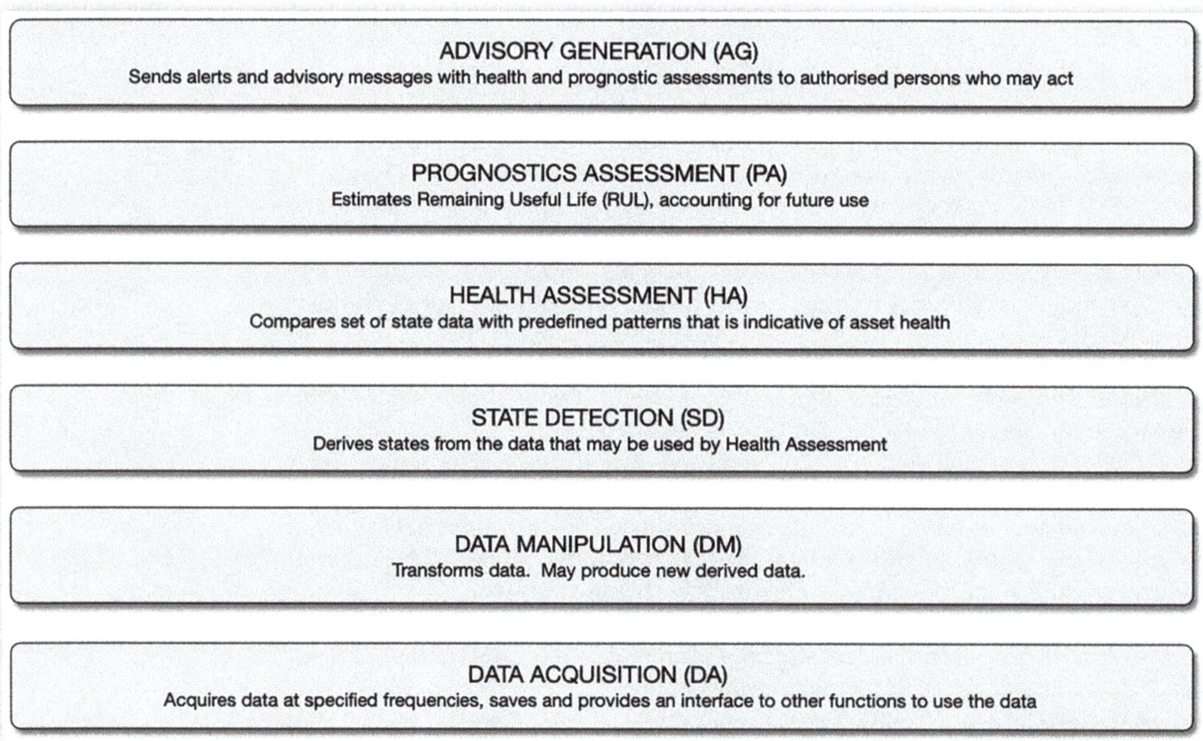

If IVHM is taken to its next logical level, it should extend the principles from the individual platform or vehicle level to an Integrated Fleet Health Management (IFHM) approach. The integrated fleet approach integrates data from all similar platforms with similar operating contexts and environments, increasing the size of data populations and improving statistical certainty. A greater number of insights in machinery behavior may be derived from a fleet view. Many manufacturers in the Aerospace market have embraced fleet wide PdM, where they can increase the statistical certainty of their analysis with access to a bigger population of assets across many customers. What insights are gleaned from the fleet view can be applied to small operators with a limited number of assets.

IFHM has considerable hurdles to jump before it may emerge as an open concept due to the latent value and commercial sensitivity of the data from the subsystems and the sharing across platforms that may have different commercial operators. The manufacturers get access to data by offering other services wrapped around their products that provide benefits to the operators for sharing data. This process of manufacturers providing services is known as servitization. Having access to the data allows a clever analyst to conduct a full dissection of organizational operations and poses a risk of loss of competitive advantages. Operators with smaller fleets also stand to gain more from this than those with larger fleets.

The adoption of IVHM where data sharing across the full product lifecycle is a goal which implies changes in how industry stakeholders collaborate and may shift industry value chains. There is much work to do to minimize disruption to value chains and profits, and attain the benefits of significant cost cutting in the whole industry.

There are a series of Aerospace Recommended Practices (ARPs) that have and are being written about IVHM associated with a lifecycle model to build an aerospace IVHM system. The general principles in the documents should be transferable to any industry. Some of the ARP documents are still being developed by the SAE Health Management (HM-1 committee). The following diagram shows the HM-1 development lifecycle and the associated ARPs that have been and which are still being produced. The production ready ARP and AIR documents may be purchased at [5]. More documents are in production and being released. AIR 6904 that covers data interoperability has been subsequently released (Figure 4.2).

What Are Maintenance Credits?

PdM has matured independently of the certified maintenance regimes in aerospace that have the proofs and quality assurance relied on to reduce risk of failure (termed Maintenance Credits). PdM has predominantly focused on failures with operational and economic consequences, reducing the commercial risks associated with servitization. It has been realized that there is an opportunity to extend PdM into being certificated, but there are numerous procedural and quality assurance challenges to overcome.

Maintenance credits are where maintenance is able to demonstrate a quantified reduction in the probability of failure, where the failure has very bad consequences. The proof that a PdM maintenance task can achieve this may rely on having access to a statistically significant set of data from many sources and quality assurance in the PdM process and applications before the maintenance credit is approved. This provides another incentive for IFHM.

This empirical evidential approach may be difficult as there is little data around failures (especially those with high impacts), because these events are rare (see the notes

FIGURE 4.2 IVHM development lifecycle and associated SAE ARP documents (Source SAE ARP 6803) Correct at December 2018.

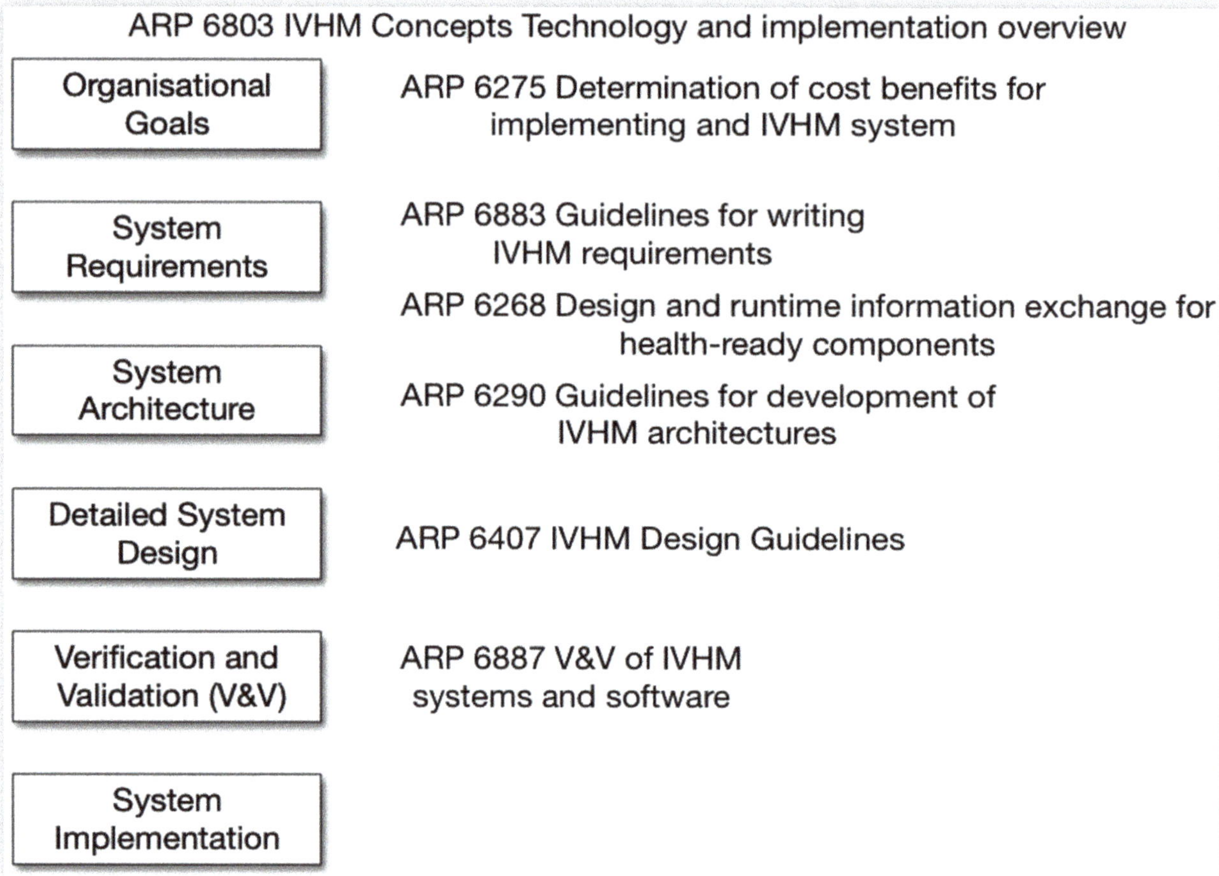

on the Resnikoff Conundrum in Chapter 9), so other approaches may need to be taken to provide sufficient evidence. These include:

- Using component as-found condition data from repair and overhaul facilities (R&O) to report the condition of components at an operating age after the components have been removed, coupled with previous time-series data from the component in-service time. Marrying up "as found" condition data to previous operating history and sensor data bolsters prognostic certainty.
- Using any post maintenance test data (such as testbed data). This data may also be used to seed other data-derived models used in PdM, because the baseline of normality will have shifted due to maintenance intervention.
- Using the data from any samples of replaced parts subsequently tested to destruction and the data associated with the test, as-found condition and time-series data from the component in-service time.
- Using any in-service NDT/NDE or inspection data and merging this in PdM.
- Using high-fidelity modelling and digital twins.
- Using any relevant design data and assumptions: if reliability growth was performed then all the data from this program should be included.

The PdM system itself will require inbuilt facilities to increase assurance that it performs the process of diagnosis and prognosis to a set standard with guaranteed degrees of certainty. For example, failures will be detected and diagnosed 95% of the time with 95% certainty.

The author collaborated with an academic colleague contributing to a chapter in [6], on trust in IVHM systems. We cited a research project STRAPP we had conducted to enhance trust in a PdM system, by making the provenance of data, models and algorithms apparent to a PdM user. This provided evidence and an audit trail that the process for building and operating a PdM system had high quality and integrity, so that an element of gaining maintenance credits was shown.

The SAE E-32 committee produced ARP5987, "A process for utilizing aerospace propulsion health management systems for maintenance credit" has been released. This document begins the process of bridging the gap between Advisory PdM (health management) systems and MSG3.

Key Take-Away Points

1. IVHM is a developing model in aerospace but is applicable to other industries.
2. PdM is integral to IVHM.
3. IVHM aligns with other open standards that help define and provide guidance for building a PdM system in the context of IVHM and maintenance management.
4. The standards have yet to properly cover operating PdM and an IVHM system.
5. IVHM is being developed to enable PdM to be used in mainstream aerospace maintenance to a high level of integrity to deliver maintenance credits.
6. IVHM should be extended to the next level, it should be Integrated Fleet Health Management, or even Integrated Portfolio Health Management.

References

1. Jennings, I.K., *Integrated Vehicle Health Management Perspectives on an Emerging Field* (Warrendale: SAE International, 2011).
2. Mimosa Standards, "Open Standards for Physical Asset Management," 2019, https://www.mimosa.org/, accessed February 2018.
3. ISO Standards, "ISO 13374-1, Condition Monitoring & Diagnostics General Guidelines," 2003, https://www.iso.org/standard/21832.html, accessed August 2019.
4. OPC Foundation, "The Industrial Interoperability Standard," 2019, https://opcfoundation.org, accessed August 2019.
5. SAE, "SAE Standards Index Page," 2019, http://store.sae.org/stanindx.htm, accessed August 2019.
6. Jennings, I.K., *Integrated Vehicle Health Management, Implementation and Lessons Learned* (Warrendale: SAE International, 2015), Chapter 3 Trust.

CHAPTER 5

Why Is PdM Generally Better than Traditional Maintenance? (How to Build a Business Case)

This chapter will discuss why PdM is a better choice for replacing other traditional maintenance if conditions allow it, and how it maximizes the utilization of all the economic life of components. Applying PdM implies doing maintenance only when it is needed, just before components are going to fail. This aligns with the policy of "*if it ain't broke, don't fix it.*"

On-condition maintenance is in principle applicable to any of the six patterns of failure (see Figure 2.5) as long as the other pre-requisites outlined in Chapter 3 are met. The underlying assumption is that the PdM on-condition is practical, effective, and cost beneficial. Other traditional maintenance tasks, such as scheduled restoration or replacement (replacement is sometimes called discard) are only applicable to failures that align with a wear-out, age pattern: these tasks cannot be applied to other patterns. Weibull analysis provides an indication of whether a part's functional failure (or failure mode) is wear-out or age related as its shape parameter is greater than 1. There is no hard boundary for shape indicating wear-out, but a value greater than 2 is a greater wear-out indicator than a shape value of 1.1, which would be very close to random.

Why Choose On-Condition over Scheduled Discard/Replacement or Restoration?

Scheduled restoration or discard maintenance organize the interventions to a time before the probability of failure increases to unacceptable levels, meaning the failure modes fit a wear-out or age pattern. The benefit is that this early intervention prevents unexpected failure and the accompanying operational disruption, avoiding, possible secondary damage.

The age for conducting scheduled restoration or replacement tasks is selected so it is triggered before the probability of failure substantially increases, and this can be demonstrated by using the results from a Weibull analysis.

If a graph of a Weibull cumulative distribution function (see Figure 5.1 below) is drawn, one can see the cumulative probability of failure at an age for a population of components. For a wear-out failure pattern, the shape of the cumulative loci is a classic "S" shape. Here there is an early period of useful life, followed by an increase in the probability of failure observed by the increasing slope in the middle of the line. This graph is a representation of probability that a component has failed at an operating age. The graph is produced by estimating the fit of a population of ages at the time of failure events, for a particular component from a single failure mode to a Weibull distribution. The cumulative probability distribution is then plotted

In the chart below a line is arbitrarily drawn for scheduled replacement maintenance to be done at an operating age that coincides with the B-10 measure (when probability of failure is 0.1% or 10%). It may be observed that the majority of economic life of the components surviving beyond the B-10 figure is being deliberately discarded as those parts are replaced. At B-10, 90% of the components are likely to have survived when they are replaced regardless of their condition. This implies that if scheduled replacement or restoration are the only effective planned maintenance then there is no known means of measuring condition. The belief that there is no means of measuring condition or detecting symptoms of failure could and should be challenged.

If a high-integrity PdM task can be applied with high confidence that the PdM will be successful, then the vast majority of incipient failure events can be detected just before they fail. This implies that there is a means of measuring condition for using PdM. The time of incipient failure detection will be at the latest possible time after all of the economic life of the component has been consumed. As long as the P-F interval is

FIGURE 5.1 Typical periodicity for restorative or replacement maintenance seen using a Weibull Cumulative Density Function (CDF) chart.

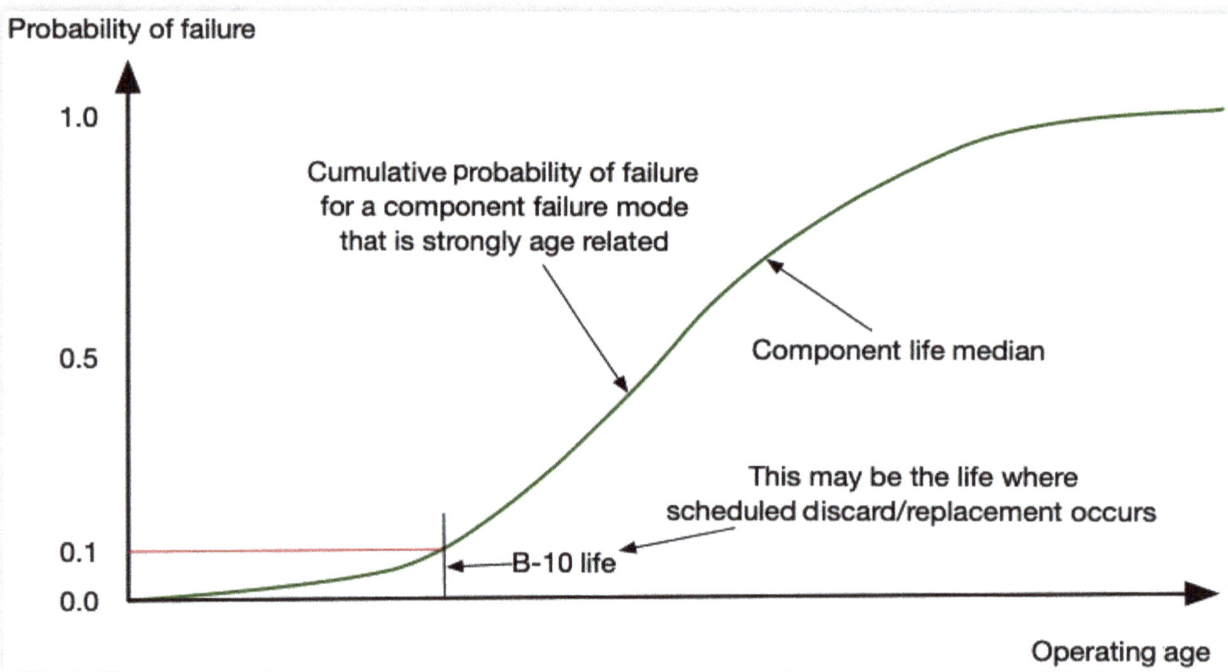

sufficiently long to regard the recovery as a planned event, then there are significant economic advantages. Through life costs have potential to be reduced by tens of percentage points for age-related failures. The following figure helps illustrate this.

In Figure 5.1, an imaginary component with a dominant age-related failure mode is illustrated. The graph shows the cumulative density function (CDF) and the increasing probability of failure over operating time (or operating cycles). It is assumed that there is high confidence that the CDF is accurate. The graph also shows the "B-10" life of the component. This is the age where 10% of a population of components is likely to have failed. It can easily be translated to a percentage by multiplying by 100, because probability is a measure in the range zero to one. In other texts (such as Moubray) this is called the "useful life" before the rate of failure increases to unacceptable levels.

It is reasonable to select the B-10 life as the periodicity for changing used parts for new (scheduled discharge or replacement maintenance task) where the Weibull shape parameter is > 2.0 (showing wear-out). If there is population of 100 components, it is probable 10 are likely to have failed by the time the discard maintenance is triggered, and the remaining 90 are replaced at the B-10 age.

The reason this discard maintenance is done is that it is trying to avoid unplanned failure, because the consequences (including the disruption) of the unplanned event are unacceptable. It is noteworthy that the discard trigger point still accepts that a small proportion of the failures will be unexpected because they lie in the left-hand tail of the distribution.

If the economic life is discarded by changing the surviving 90 components before they have failed, it is possible to see by eye from the graph that the waste of usable economic life of the discarded components is considerable.

If there was a regime that took away any preventative maintenance and allowed the population of components to run to failure, then the median of all the ages at time of failure would be the best "expected" age for this component. It is axiomatic that the median age is much greater than the B-10 age.

If it was possible to apply fully effective on-condition maintenance,

- The system detects every single potential failure
- The system does not signal detection of a potential failure where none exists (avoids false positives)
- The P-F is sufficient to plan recovery and minimize operational disruption

If on-condition maintenance prerequisites are met, and on condition is effective it means organizations could utilize all possible economic lives of the components, minus the relatively very short P-F interval. Effectively, PdM is enabling the economic utilization up to their median lives for the population of components. The perfect PdM solution also eliminates the unplanned failures accepted by the scheduled replacement regime (the likely unplanned failures of 10% of the population up to the B-10 age).

It is axiomatic that the median value always exceeds B-10, and therefore value is generated. The value is (with perfect PdM):

- Full utilization of all possible economic life of the components, the material costs of components are consumed over a much greater period of utilization time.
- Elimination of unplanned disruptive failures, all potential failures have enough P-F intervals to plan the recoveries at a time minimising operational disruption.
- Reduced frequency of maintenance work, this will include reductions in
    - Labor cost.

  - Outage time (although one component might not change the normal planned outage time cadences and durations). Unplanned outage time will be reduced.
  - Material costs.
- If the reduction of unplanned outage is reduced, then that time may be spent in profitable utilization of the asset generating more revenue or higher availability rates.

In the real world, PdM systems are rarely perfect, diagnosis will have errors, some failures will be missed (false-negative), and sometimes the PdM system reports detect failure where no failure exists (false-positive). The performance of the PdM system must be optimized to reduce these errors so that the benefits of using PdM are realized. The difference in value between scheduled replacement/restore and PdM is so great, that a reasonably small level of PdM errors is still acceptable and will still deliver considerable value.

Some terminology used in statistics may also be useful to know. The ability to tune a diagnostic to capture all of the genuine failure events is optimising the systems "sensitivity" which addresses false negatives. Increasing sensitivity may trade off "specificity" where you accept higher false-positive rate. The PdM tuning may be a continuous process in order to maximize benefits. The language of sensitivity, false-negatives, specificity and false positives are associated with confusion matrixes described in greater detail elsewhere in this book.

Another factor that supports the use of PdM is that it should be cheaper to apply compared with traditional on-condition maintenance. A PdM system should be highly automated and able to continuously monitor, allowing a user maintainer to monitor a huge number of components with the system alerting the user when they need to take action. The costs of applying highly automated PdM compared to using traditional on-condition maintenance using periodic manual surveys is a fraction of the latter. In the Rolls-Royce PdM system in the early days an expert could manage up to 100 assets, as automation developed over the years this figure grew to thousands of assets with vastly improved automated diagnostics and prognostics.

The combination of cheaper PdM through automation, greater utilization of the whole of the economic life of the component and reducing the frequency of replacing the part, the benefit of applying PdM over traditional scheduled restorative or replacement maintenance is considerable. Therefore, in the RCM system, on-condition maintenance is preferred over scheduled restoration or scheduled replacement tasks.

Exceptions to the Rule — When Is Scheduled Replacement Better?

When thinking about applying on-condition maintenance, other factors may play a part. It is worth considering these when judging the effectiveness of the on-condition tasks.

Some component failures have minimal consequences and it might not be cost effective to conduct preventative maintenance. These should be designated as 'No Scheduled Maintenance' (NSM) and allowed to run to failure (RCM: Chapter 6).

Another case where PdM may not be the optimal maintenance choice is where many the same components exist and have a very high degree of redundancy, but each requires a lot of effort to individually change. An example is factory lighting where the lamps have a wear-out failure pattern. Changing a single lamp may be expensive (if scaffolding needs to be erected, disrupting production lines, etc.), so there is no advantage in condition monitoring leading to individual lamps being changed: the most effective maintenance task to apply is scheduled replacement of many lamps as possible in one task (spreading the cost of erecting the scaffolding and minimising loss of production) regardless of the condition of the lamps. A variant of on-condition is to wait until a number of adjacent lamps have failed, reducing localized lighting levels to a just acceptable level, and then initiate immediate batch changing. This alternative strategy has a risk of contravening health and safety (unacceptable lighting in a workplace) legislation, and so must have safeguards to avoid breaches.

When Should Median and Mean Measures Be Used?

In this book the author has used median over mean to measure a distribution's expectation measure (a measure of its "middling tendency"). This section explains why.

If the median life of a population of components is calculated, it will show a summary measure of the "expected life." Within statistics the "expected value" equates to the best measure of the middle tendency of the data. Why choose the median? Why not use mean or average (or in other words the Mean Time Before/To Failure (MTBF or MTTF))?

The median is selected for two reasons:

1. Because it is a better measurement of middling tendency or expectation of continuous data if the underlying distribution is not uniform and symmetrical like a Gaussian bell-curve. With a symmetric normal or Gaussian distribution, the mean is the same as the median.
2. The mean measure is also far more sensitive to outliers than the median measure. Sometimes statisticians use "trimmed mean" to eliminate the undesirable effects of outliers. The author merely prefers using the median.

There are other reasons why MTB(T)F should not be used:

Merely quoting MTBF with no other information for equipment reliability means some assumptions must be made. These are:

- The underlying failure rate is constant.
- If the underlying failure rate is constant, then the probability distribution that applies, that the failures best fit into is the Negative Exponential Distribution. Not a normal distribution.
- The mean for this distribution coincides with a probability of 0.63, which translates as 63% of a population are likely to have failed at the mean age (probabilities are always expressed as a number between zero and one).

Many lay people may expect the mean would equate to 50% (which would be correct if the distribution were normal or Gaussian). The assumption that the failure rate is constant is also a special case in the real world. Reliability is defined as "The ability of an asset to continue functioning, for a stated time, in a stated operating context". Using MTBF as a metric risk vastly over-simplifies what reliability actually is and should be used with great caution.

The point is without knowing any other information, the MTBF does not explicitly say how many assets have failed at the MTBF. It is therefore highly questionable then that MTBF is actually a true measure of reliability. The author personally tends not to use MTBF and relies more on Crow-AMSAA and Weibull (sometimes other distributions) to understand reliability.

Understanding the implications of Weibull distributions and how they may be estimated using failure (and censored removal) data and how scheduled replacement or on-condition (PdM) maintenance works with respect to the Weibull, enables the calculation of costs and benefits using simple simulation.

Why Is the Use of MTBF Persisted?

The persistence of MTBF (MTTF) is because it has always been done this way and the metric is deeply embedded in society. The mathematics associated with MTBF reliability calculations are simple. This means it is easy to teach to non-technical people in courses that only have time to cover rudiments of the subject. Its use as a reliability metric is highly dubious and its use to try and calculate periodicities for scheduled maintenance must never be done. The RCM process covered in Chapter 6 has references for how to properly calculate maintenance periodicities for different types of maintenance tasks. Many people are not taught the basics of statistics, including many engineers. In these modern times maintenance managers will be expected to make decisions and act on data presented to them. They need to understand how the data has been treated statistically to ensure it is not misleading.

Building the Business Case

Building a business case for PdM is difficult for two reasons. The first is that PdM, to date, is similar to taking out an insurance policy: you pay up front to mitigate unwanted events that may or may not happen. If PdM is only regarded as insurance, it makes the investment discretionary and less likely to be approved against other spend in the organization. Secondly, PdM is probabilistic and not deterministic: diagnosing functional failure may miss genuine failure events or may diagnose failure where none exists (false positive), prognosis has uncertainty and may fail to forecast Remaining Useful Life (RUL) accurately. Most business cases are mainly built on deterministic assumptions such as a constant failure rate with no variation. Some may have worst, medium, and best-case assumptions, but even these are built deterministically. Although well intentioned and easy to grasp, they may lead an organization to make grave mistakes. A business case built on probability, variation and simulation would be harder to sell to business-orientated people who may be unfamiliar with probability.

The difficulty of justifying the investment in PdM and making a convincing case for PdM based on probability might be overcome by using simulation to avoid the difficult mathematics of a dynamic system. Simulation can be used to provide statistically significant data that builds confidence that the results may be trusted. Simulations can be built with dynamics shown graphically which helps the business users understand the system and assists in building trust in the results.

Business Cases Built on Reliability-Availability-Maintainability (RAM) Simulation

Reliability–Availability–Maintainability (RAM) modelling is the term used in the US for what is often called ARM modelling in the UK. These models consist of utilizing Weibull distributions and knowing the periodicity of Planned Maintenance using a discrete-event Monte-Carlo simulation to simulate the running, failure, maintenance, and recovery of fleets of assets in order to determine their availability and reliability. The RAM model may then be extended to include cost data (loss of production while unavailable, maintenance labor and material cost for either planned or corrective maintenance, etc.), to build a cost model to contribute to a trustable through-life cost calculator. This type of model can then be changed to run "what if" scenarios such as changing maintenance from "planned replacement" to PdM to determine the cost benefits of doing so. These simulation models will be easy to build using Python and some of its specialized libraries [1, 2]. Many business cases may be built using static spreadsheets where failure is represented by MTBF with their assumed constant failure rates and using perfect maintenance plans (that are not disrupted by unexpected failures). This is most often not representative of real-world conditions and will result in an inappropriate budget allocation. This is likely to result in undesirable emergent behavior where the greatest efforts of the maintenance team will be spent on meeting inappropriate budgets as they struggle with containing failures.

PdM has other considerable benefits because it gathers much data, and the behavior and condition of machinery are known. In a situation where corrective maintenance needs to be conducted, PdM can inform the maintenance planner whether other components that have been made accessible by the corrective maintenance are near the end of their lives and whether it is economic to change these components now. This can discard some residual economic life but also mitigates the risk of failures before the next planned outage (or planned maintenance period) of the asset. The basic idea for looking for opportunistic maintenance is increasing the probability that the asset will not have an unexpected breakdown before the next planned maintenance outage.

The second improvement is that greater data is available to help the RCA process where investigations into reliability may be targeted better (PdM provides better data on failures), and can help isolate the preventable causes of failure.

The elements needed in a business case run by a RAM model are:

1. The Weibull shape and scale parameters for each component's failure mode so that the simulation can randomly sample this to generate simulated failure events
2. The cost of sampling, or inspecting. For a PdM system, the cost of acquiring, transferring and analysing data in a highly automated system will be low. The costs of manual inspection may be higher
3. The costs involved with not detecting a failure, which may include lost production time (disruption to schedule), and any extra labour or materials needed to recover from a complete failure
4. The labour, material and possible lost production time for a planned recovery, when the on-condition works. Lost production might not be included if recovery is included within a planned outage.
5. The on-condition effectiveness of
   a. probability of detecting all simulated failures
   b. The false positive rate where nugatory planned recovery maintenance is carried out

The RAM simulation is run over a period of selected time (maybe the full life of the asset), thousands of times and summary statistics used to calculate the distributions of the costs. These can be used to determine upper, median and lower bounds of cost at several pints in the assets lifetime.

Key Take-Away Points

1. If on-condition and PdM is practical and cost-effective with minimized false positive rate it is most often a better choice economically than other traditional maintenance tasks (scheduled restoration or replacement).
2. Business cases should be built using Weibull metrics on RAM discrete-event simulations. These simulations are relatively easy to build with the advent of Python and its specialist libraries.
3. PdM helps the build of knowledge of how machinery operates and degrades to a far better degree than without it.
4. Using MTBF for reliability analysis is an oversimplification that can lead to incorrect management decision making. MTBF is arguably not a measure of reliability.
5. PdM data and knowledge should be used in Root Cause Analysis (RCA) processes.
6. PdM and prognostics should be used as intelligence to conduct opportunistic maintenance when other preventative or corrective maintenance is necessary because of a breakdown, due to knowing the likelihood of failure and condition of other components that could also be worked on.

References

1. Simpy Developers, "Discrete Event Simulation for Python," 2019, https://simpy.readthedocs.io/en/latest/index.html, accessed July 2019.
2. Salabim Github Repository, "Discrete Event Simulation for Python," 2019, https://github.com/salabim/salabim, accessed July 2019.

CHAPTER 6

How Does PdM Relate to Reliability-Centered Maintenance (RCM)?

The aerospace version of RCM is known as MSG3. MSG3 and RCM and both share a common ancestry back to the original RCM report [1] conducted by United Airlines.

On a personal note, the author has spent most of his professional engineering working life engaged in operating and sustaining in-service complex assets. There is a perceived schism between design and in-service engineering, where it seems more prestigious to be a design engineer. Many of the engineering frameworks, such as Systems Engineering, Reliability Engineering, and IVHM, originate in the design stage of the complex asset lifecycle. RCM originates in the in-service phase and values the knowledge of experienced operators and maintainers above the knowledge of the design engineers while recognizing design knowledge is valuable. RCM is the voice of the in-service engineering discipline.

There is much marketing material that claims PdM is superior to RCM or supersedes it. This chapter will explain why this is not the case, that PdM is complementary to RCM and the two frameworks are not in competition. PdM does not supersede, nor is better than MSG3. PdM may be included in MSG3-derived maintenance regimes in aerospace if and when the methods for IVHM/PdM maintenance credits are resolved.

The seminal Moubray book RCM II [2], has recently been revised with the release of "RCM3, risk-based RCM" [3] where the process has been updated to answer eight basic questions, from the previous version's seven questions. This book is fully aligned with these updated principles.

RCM 3 requires these questions are addressed

1. *What are the operating conditions (How is the equipment or system being used)?*
2. *What are the functions and associated performance standards of the asset in its current operating context?*
3. *In what ways does it fail to fulfil its functions (failed states)?*

4. *What causes each failed state (failure modes and failure mechanisms)?*
5. *What happens when each failure occurs (failure effects and consequent severity)?*
6. *What are the risks associated with each failure (inherent risk qualified)?*
7. *What MUST be done to reduce intolerable risks to a tolerable level (using proactive risk management strategies)?*
8. *What CAN be done to reduce or manage tolerable risks in a cost-effective way?*

The first six questions RCM poses can be answered within a maintenance-centric Failure Modes and Effects Analysis (FMEA). The answers to questions seven and eight may be derived using an RCM decision tree (see Figure 6.3 or the MSG3 version in Figure 2.5). The FMEA is a widely used tool to determine how items may fail, their risk and how critical the failures are, with an explanation of consequences. The FMEA is described in more detail below.

There are many claims that PdM supersedes or is an improvement on RCM (or MSG3 in aerospace). The comparison does not make sense. RCM is a framework that defines, builds, and sustains a maintenance system or regime for any asset, whereas PdM is an advanced type of on-condition based maintenance taking advantage of digitization, continuous sensor data, and automated data analysis. One of the outcomes of RCM is to select the most appropriate maintenance tasks to prevent or avoid loss of function: that includes on-condition predictive maintenance tasks. The tasks are selected if they are effective and efficient. PdM may be specified as a natural consequence of running an RCM process, and this makes it obvious that PdM does not supersede or replace it. Within the RCM process the PdM should be selected if the right knowledge is held about PdM and how it works.

RCM is a rigorous process that was developed for industry that has gained an undeserved reputation of being too onerous, bureaucratic, and often wasteful. There are many other "streamlined" or alternative RCM processes that have been pitched that have often missed out several vital elements of the analysis, occasionally leading to inappropriate maintenance that could be dangerous. This was recognized by John Moubray the originator of Industrial RCM who, with others, developed SAE standards JA 1011 [4] and later JA 1012 [5] which provided a list of elements that must be in a process that claims to be RCM compliant. Moubray's RCM 2 book [2] is a key reference used by these standards. The RCM process needs to be done properly and will pay back handsomely when done.

One of the keys to writing useful FMEAs or using them is predicated on some of the information that needs to be captured within them. It is known that PdM systems utilize sensor data that can be continuously sampled. Additionally, PdM can also take account of other condition data where samples are taken periodically (discussed in the principles of IVHM).

The FMEA is usually arranged as a grid of columns and rows (a table) with text entered in each of the cells. There are many formats of FMEA, and sometimes in large organizations these formats are predetermined. Many formats are not as useful for designing a maintenance regime, including PdM. The following section defines the column headings used in an ideal FMEA contents that is optimal to help adoption of PdM. The column headings are in bold.

1. **The FMEA row reference number**. This uniquely identifies the main item in the next column. The numbering system may use an indenture system where the identifier is split to identify the level of indenture that the equipment item exists in a bill of material.
2. **The equipment items**. A description of the item and any identifiers that may be used for the physical position or identification to the item in the wider asset.

3. **The equipment item functions.** Functions should be split to show whether they are primary or secondary functions. An equipment item may have many functions. Functions may be shared between equipment or a lower level equipment's primary function may be a secondary function of a higher-level function. There is a one-too-many relationships between the equipment item and its functions. Some functions may be shared between items of equipment. Functions need to be expressed with required performance bounds or limits. This is extremely important in understanding the difference between normal and abnormal behavior.
4. **Functional failures**. This follows on from functions as a description of how functions may not be delivered. Once functions are identified the functional failures are usually very easy to define.
5. **Failure Modes**. The failure modes that result in functional failure described above.
6. **Failure mechanisms**. An FMEA that includes failure mechanisms is often called FMMEAs. Where the knowledge exists, it is well to include this information, because it may allow organizations to take action to avoid the failure modes, and it may help the selection of sensors to apply for on-condition maintenance. In this way, including failure mechanisms assists the adoption of PdM.
7. **Failure causes.** From one perspective, failure mechanisms are the root physical causes of failure. This entry is still required because other predominant causes are people and quality. Failure may be caused by overloading or operating plant outside its design intent. Bad maintenance (possibly due to lack of training) may also be a cause of failure. It is often helpful to think in terms of credible hazards or vulnerabilities associated with the environment processes and organization to properly populate this section. The FMEA may extend its usefulness if any data that is discovered in any subsequent Root Cause Analysis (RCA) process is back populated in the FMEA.
8. **Effects and consequences**. It may be necessary to break down different effects for different parts of the mission profile, or variation in the environment (winter/summer):

    a. **Immediate local effects.** What happens to the part itself when it fails?

    b. **Immediate higher-level effects**. What happens to the part and dependent or adjacent equipment? This may be effects to the part's parent assembly or system.

    c. **Immediate whole-asset level effects**. What happens to the whole asset or train of equipment in an industrial setting? Is the whole asset made unavailable or not? If so, this will have significant operational impact.

    d. **Higher level consequences**. If the failure disrupts operations or reduces the quality of the product being produced (in the industrial context).

    e. **Business effects**. Effects on reputation, meeting legislation, avoiding fines etc.

    f. **Recovery or corrective maintenance**. How long, when, what resources are needed (human and material), and what is the likely cost. Information about recovery is part of prognostic advice that should be given with Remaining Useful Life (RUL).

    Effects severity may also be categorised as SHEL (Safety, Health, Environment, Legislation), Operational, Economic, Insignificant, see the explanation below.

9. **Likelihood or probability of occurrence.** Effects severity may also be categorised as SHEL (safety, Health, Environment, Legislation) Operational Economic Insignificant see the explanation below. These are traditionally

provided as an ordinal value at the most basic of "infrequent," "normal," or "frequent" that may be quantified as a probability of failure within a time unit. For example, each asset having one failure per year.

10. **The severity (or criticality) of the impact.** This is where it is possible to classify the failure as per the breakdown shown earlier on the Y scale of Figure 3.8. This will be further explained later.
11. **The detectability of the failure.** This section is especially important to applying PdM, because it provides baseline information that can be used in the RCM decision logic to answer the question, "Is on-condition (or PdM) practical and cost effective?"
    a. **Is the failure "hidden" from the operators or maintainers?** Were there any alarms or warnings or other indications or symptoms present before or during the failure? Is a normally hidden failure become apparent in other abnormal machine states (e.g., a stuck pressure relief valve fails to open when an abnormal overpressure excursion occurs)? Capturing the time these symptoms presented themselves before failure is useful data to support prognostics.
    b. **What is the nature of the failure?** Was the failure a gradual deterioration or was it intermittent or sudden/catastrophic? On-condition maintenance relies on having a time gap between failure initiation and final functional failure. If the failure is sudden there might be no practical P-F interval. This time delay is a pre-requisite for applying on-condition maintenance.
    c. **What sensors are available that may be used to detect failure?** This obviously plays directly into applying PdM. Inferring how symptoms may be observed from seeming unrelated sensors is also an intellectual puzzle most engineers will relish.
12. **Any dormancy in detectability**. Can the equipment operator or the maintainer undertaking their normal duties detect the onset, progress, or failed state? This is important to identify hidden failures that are considered in their own branch of the RCM decision logic.
13. **The Probabilistic Risk Number (PRN)** or **the Criticality.** A combination of:
    a. Likelihood
    b. Severity/impact
    c. Detectability (of the failure)

    These methods of determining criticality are described in more detail below.
14. **The level of risk:** This is using similar impact and likelihood measures that are used to determine criticality, but other business-orientated views may be taken around risks.

It is easier to estimate impact or consequences of failure than it is to determine the likelihood or frequency. Care need to be taken when working out the criticality of high-impact but low-likelihood failure modes. It may be better to increase the criticality for these items to compensate for the lack of accuracy in assessing the likelihood.

How Can Severity of Failure Be Categorized?

A breakdown of classes of impact of failure was introduced in Figure 3.8. This breakdown of classes split severity into 4 groups. It is worth developing this idea here in order to classify FMEA effects and consequences in greater detail.

The break-down of severity is as follows:

Safety, Health, Environmental, and Legislative Compliance (SHEL). This is the most severe classification, and repeated failures of this nature threaten the existence of any organization that owns or uses complex physical assets. It is possible to break this category down further to capture where safety and health issues cause death, or lower down the scale whether debilitating injuries are caused. Environmental issues are also unsustainable with restrictions on emissions becoming ever more stringent. PdM can be used to monitor emissions and provide information that enables management and minimization. Legislation is tied up with SHEL but may also be associated with other operational restrictions that need to be taken into account. PdM can be utilized to monitor these aspects and report compliance if regulations, standards, or laws require this. A severity in the SHEL classification automatically has operational and economic impacts.

Operational. This severity impacts the ability to operate a complete asset that compromises production, disrupts operational schedules, or product quality. This has economic impact as well. Product quality is related to the quality of the output of the machinery being monitored. This may be more or less important dependent on the context of the equipment and how it is used.

Economic. This severity is where avoidable cost is incurred that has no operational impact. The cost of maintenance material or recovery may be high. This can apply to the replacement of intrinsically expensive parts that have little impact on operations.

Insignificant. This may seem obvious, but it is important to know because one needs to prioritize and focus effort on. There is a danger in PdM that a purely data-driven approach that lacks domain expertise may become trapped in dealing with insignificant parts because the data is so rich. It is necessary to positively know what equipment failure is insignificant in severity so it can be actively avoided in consuming resources in low-value PdM analysis. This also provides a defendable justification for why maintenance is not done.

How Can the Likelihood of Failure Be Categorized?

The likelihood of failure may be categorized in several classes from extremely unlikely to frequent. What this means depends on the context of the machinery and what has been historically accepted or not.

The Applicability of Weibull Analysis

One useful tool is the Weibull analysis covered in Chapter 3 with type 1 prognostics. This allows us to express likelihood of failure as a probability distribution that is much more satisfactory than a figure like Mean Time To/Between Failure (MTBF). A Weibull distribution can be defined using two variables: the shape and the scale. The shape determines the shape of the distribution varying between 0 to 1 (that would indicate infant mortality), 1 that indicates random (the probability of failure is constant over age), or greater than 1 where the probability of failure shows wear-out. The scale is a measure of the age where the probability of failure is 0.63. In other words, the age by when 63% of a population of parts is likely to have failed.

A useful factor of considering what maintenance type tasks in an RCM process is based on the shape parameter.

- If shape is < 1, then the maintenance intervention is discovering the root cause of the failures (as it is a quality issue) and avoiding the cause.

- If shape = 1, the failure pattern is random, the only applicable maintenance task is on-condition (or PdM).
- If shape is > 1 (but more usefully > 2) then on-condition or scheduled restoration or replacement tasks are applicable. On-condition is preferable to the other types of maintenance.

Where Weibull analysis may be useful in terms of a FMEA is to use what is known as the B-20 age. B-20 is the age at which 20% of a population of parts is likely to have failed given a Weibull distribution. This measure is low enough to be sensitive to infant mortality or wear-out characteristics. The FMEA could have a range of bins that the B-20 would fall into, thereby providing a machine-centric range. For example: the FMEA bins might be between 0 and 1000 h, 1000 and 2000 h, 2000 and 3000 h, etc.

It is also worth noting that B-20 is a measure of reliability and may be more sutable to use compared to MTBF.

FMEA Storage and Tools for Analysis

The initial job of defining an FMEA is to work out a physical breakdown of an asset undergoing an analysis. For example, an aircraft breaks down into systems such as avionics, propulsion, structure, etc. which then breaks down further to reach levels of parts that correspond to being functionally significant or constitute a maintainable part. The data structure for this is a hierarchy of physical components that break down in distinct levels (sometimes called indenture levels) into ever simpler constituent parts. For example, an aircraft breaks down into propulsion, structure, and avionics at the next lower indenture level. In systems engineering, system diagrams are often broken down into ever simpler levels of a similar hierarchical structure. A similar hierarchy based on functions is also possible with asset-level functions on the top, breaking down into ever simpler functions. The two hierarchies, physical and functional, are related. A physical component fulfils functions. A function may be delivered by physical components. This means that there is a many-to-many relationship between functions and parts.

A many-to-many relationship makes it extremely difficult to ensure its integrity using tools like spreadsheets: other data management tools are required.

The traditional data structure for an FMEA is to record it in a table. This has the advantage of being understandable for humans to read and review. It does, however, have disadvantages if an organization wants to use electronic means to help analyze the data. Many organizations produce tools that allow the FMEA data to be stored in spreadsheets or a relational database that supports relating the data attributes. If the data is stored in a relational database, the data may then be queried using the SQL (Structured Query Language) facilities of the database. SQL is far more powerful than filtering and searching using spreadsheets. Relational databases record data in separate tables of data resembling smaller spreadsheets. Each table may be related to others by including primary and foreign keys (unique, identifying data fields in the table) to explicitly link the tables. The degree of relationships can be one-to-one (a person has one brain) or one-to-many (a person has many fingers), allowing a rich representation of real-world attributes. The traditional FMEA table type grid may be reproduced as a report output from the database, retaining the advantages of humans finding the table format easy to read and review. There are free, open-source, database-management systems available see (Ch 6, Reference 5). These would allow any organization to build their own database for an FMEA.

A more powerful way of structuring FMEA data that is much rarer than using the relational model, is as a graph data structure. Why would we do this? The FMEA data is very rich in attributes (like different levels of impact of failure) that relate to each other. The key factor is that many of these attributes may be shared between many of the other entities in the FMEA. This means it is very difficult to remember that a common attribute such as "local impact" to a number of functional failures might be entered slightly differently using a spreadsheet with free-flow text entries. A way to address this is to use drop-down lists with a selection of predefined entries. Using drop-down lists also have disadvantages. If the list is too short, the choice of entries do not cover the real world situation the user is trying to record. If the list is too long, the authors own experience suggests the limit should be about 20 entries. Then the user often does not spend the time necessary to pick the most appropriate item. Data integrity is often degraded because users may select the top item in the list if they are unsure. Another situation that occurs is if two or more entries are applicable. The graph data structure is optimal where the data set is rich in relationships that are "many-to-many" in nature. These many-to-many degrees of relationship between entities describe the true nature of an FMEA.

There are now free-to-use Graph Database Management systems available (Ch 6, Reference 6), along with standards used to create what is known as Linked-Data (Ch 6, References 7 and 8). In the linked model, the data exists in a series of triples that are broken down as "subject-predicate-object". Concepts such as "leak" – "is a type of" – "failure mode" can be formed in the triple structure that allows us to form a semantically rich set of relationships creating a network or nodes (subjects and objects) linked by edges (predicates) as illustrated below (Figure 6.1).

The triples can be queried using SPARQL which is the equivalent to the Structured Query Language (SQL) in a relational database specifically designed to query triple databases. One of the advantages of the triple store is that the data store, itself, can infer new relationships (predicates) by logic within the SPARQL system by applying rules.

For example, a rule may state: if Charles is the son of (Patricia and Edwin) and Peter is the son of (Patricia and Edwin) then the system may infer Charles is the brother of Peter (and vice versa). Any other siblings sharing the same parents would also inherit the same predicates (is the brother/sister of) automatically. The triple database, thereby, becomes hugely powerful for extension because of the ease of adding metadata and new predicate linkages. These automatic logical inferences are not possible using basic spreadsheets or "out of the box" relational databases. The rules could be encoded, but this would take a lot of effort.

The graph database structure is particularly suited to mining the data where relationships between the data entities is particularly rich. This is eminently suitable for data such as that contained in an FMEA. The following figure shows a possible graph linked

FIGURE 6.1 An example of a Triple used in linked data.

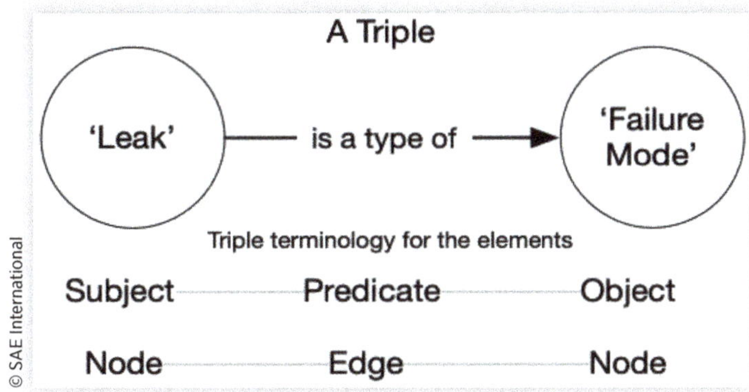

triple schema for the FMEA described in this chapter. The nodes (circles) edges (connecting directional lines) equate to the triple "subject (Node)-predicate (edge)-Object (Node)" where the object is being pointed at. Each Node represents an entity that may be decomposed into attributes. For example, the Function entity may be decomposed into primary and secondary functions, all described using triples (Function)—[decompose to]→(Primary Functions) and (Function)—[decompose to]→(Secondary Functions). Navigating through the triples is semantically rich and easy to understand what relates to what.

The author believes an FMEA would be best suited to store as a graph database. Another factor that may be relevant is the degree of the relationship. In a spreadsheet, the degree is binary, it is either true or by its absence is false. It is possible to define "degrees or strength" of relationships, but it can be overcomplicated. In a graph database this is easy to do, by including relationship attributes. For example, it would be possible to calculate the probability that a symptom was associated with a failure mode. The relationship is the association, and the unique probability can be assigned to the instance of the relationship (Figure 6.2).

In a PdM system, it is highly desirable to keep a live and accessible FMEA that can be updated during the RCM "age exploration" phase, where lessons learned and new facts may be added, as well as adaptations of the FMEA made when equipment may be modified. An FMEA should be revisited or reviewed after the following changes in an organization:

1. Standard operating procedures are changed.
2. Reductions in personnel – especially where hidden failures are concerned.

FIGURE 6.2 Graph data schema for an FMEA.

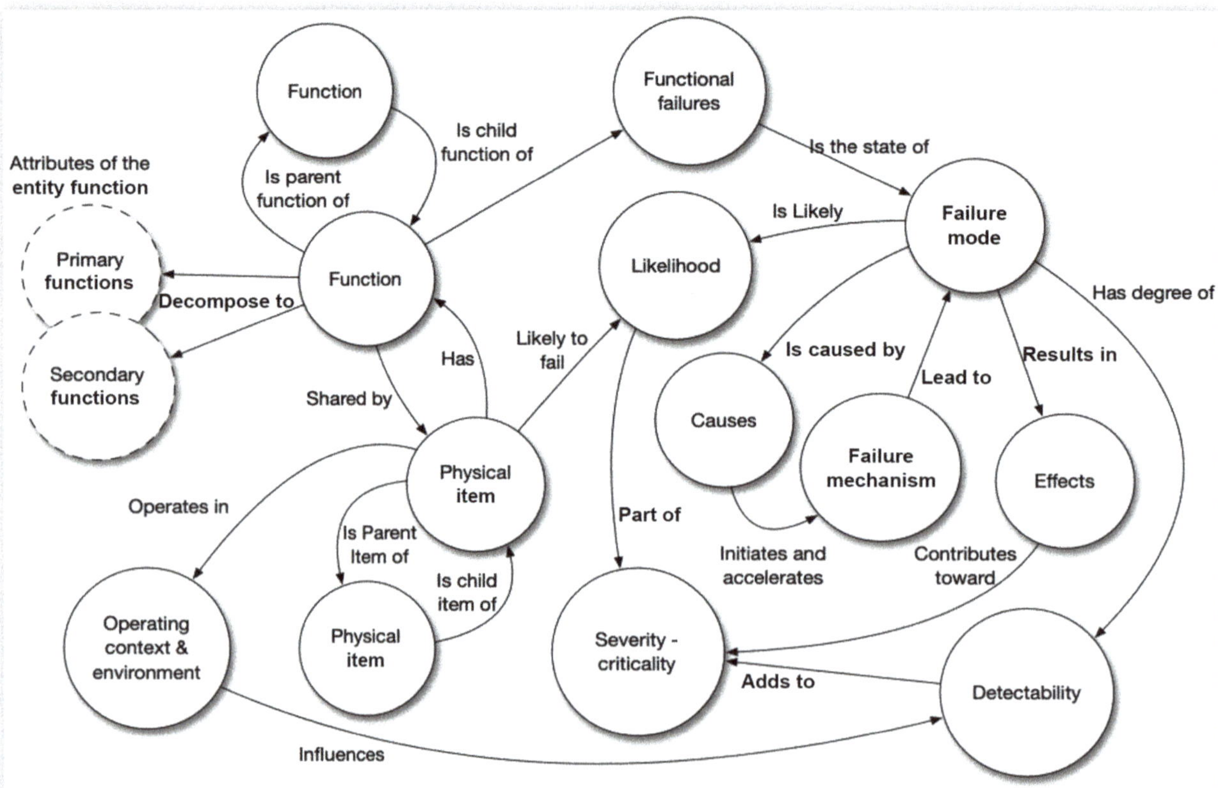

The history of application of PdM in aerospace is that it has been developed outside of the MSG regime, where the MSG preventative maintenance needs to prove that it reduces the residual (from design and manufacture) risk of failure events to acceptable levels. This validation process is expensive and detailed, and to-date PdM has not been adopted wholesale within the MSG regime.

PdM has been developed and used most effectively by some airlines and aerospace manufacturers and has been applied to supplement the existing MSG3 maintenance regimes. It also helps de-risk servitization where manufacturers offer services based around their products.

The outputs of the PdM have been advisory only, mitigating operational and economic risks and do not yet have a basis of Verification & Validation (V&V) such that the performance can deliver predictable results. This does not imply the results of applied PdM are not dependable or valuable. PdM delivers considerable business value in terms of mitigating operational and economic risks.

There are studies being undertaken by the SAE HM-1 committee (E-32 ARP5987) to write ARPs on how to achieve "maintenance credits" using PdM.

The RCM process is based on gathering data around the eight questions quoted above. The data for questions 1-6 is contained in a Failure Modes and Effects Analysis (FMEA), and an RCM decision logic is used to determine what must be done at what times. The RCM decision logic in Figure 6.3 is equivalent to that for MSG3 in Figure 2.5.

PdM is a subset of on-condition maintenance and is considered within the RCM decision logic, above other maintenance task types such as scheduled restoration or replacement. The aim is to reduce the likelihood of failure or reduce its impact to acceptable levels. The order of consideration of each of the tasks is significant: On-condition

FIGURE 6.3 Simplified RCM Decision logic.

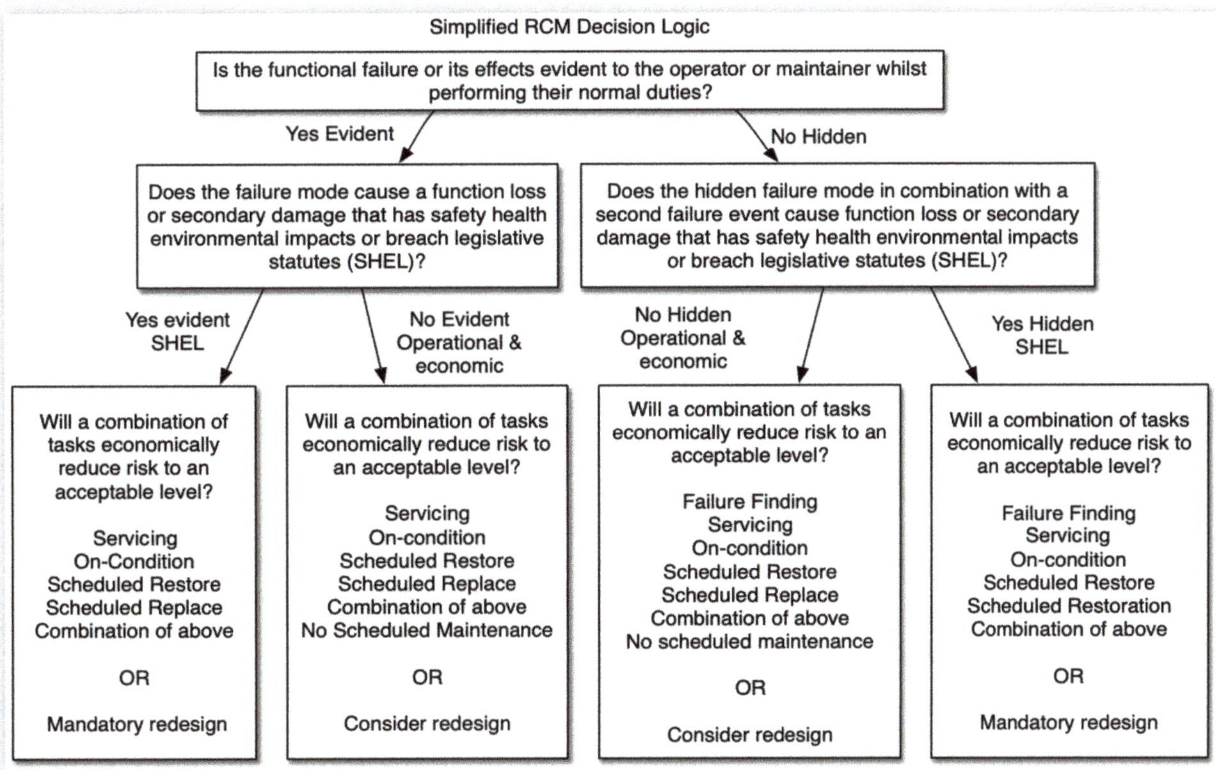

is preferable to scheduled restoration or scheduled replacement. Many RCM processes do not consider a combination of the tasks as satisfactory, which the MSG3 process does.

The scope of redesign includes changing the standard operating procedures to avoid stress (on the asset) or physical redesign of the asset itself (modification) to improve reliability or reduce the failure impact. For an operator using manufacturer's machinery, this may involve modification by the operator or working with the manufacturer to mitigate safety-related failures. Intrinsic safety of an asset is rarely an issue because the major focus of the design process of the manufacturer is eliminating safety risks, safety assurance being an existential issue. The other elements of SHEL may be dependent on the assets operating regime and environment and legislation may change during the asset's life. Modification by operators or specialist companies may supply solutions.

Of special interest is the hidden failure situation. Good design ensures that safety protection devices "fail safe," that is to an "evident" state, usually to a tripped condition. There are some devices (such as pressure relief valves) that cannot be designed this way. Many operating organizations have business cycles where they economize and save costs. One way this is done is by manpower reductions. It is vitally important that if operator or manpower reductions are made that this is done in the knowledge that this may make some failures hidden, and some failures may remain dormant when experienced people who used to walk around their machinery to inspect it have been made redundant. These inspections are not usually officially recorded in a maintenance management system and, therefore, the benefits of this normal "best engineering" practice are not apparent to accountants conducting a cost-benefit study. It would be useful to capture data where incipient failure or fault conditions have been found on these types of rounds or walk downs.

A major aspect of RCM that was extensively written about in the original Nowlan and Heap RCM report was a concept called "Age Exploration". This is where data analysis and updates to the RCM regime are embedded to update the maintenance tasks as lessons were learned. Age Exploration has not been fully developed by RCM or its aerospace equivalent MSG3 but, with the advent of PdM, the opportunity exists to develop the concepts. PdM supplies much of the data necessary to deliver age exploration and, coupled with reliability engineering (using Weibull analysis), could be further exploited as a continuous improvement process.

RCM age exploration is the systematic collection of data about parts in their operating contexts in order to establish their condition and failure related to their operating age and comparing this against their declared intrinsic reliability. Classifying which of the RCM failure patterns apply to parts is conducted. This is used to verify the RCM assumptions and original decisions to help revise the types of maintenance tasks and their periodicity selected to prevent or avoid unplanned failures. Age exploration also verifies design assumptions, although feedback of data from in-service to design teams in the product lifecycle needs to be improved. Weibull analysis described in more detail in this book would be a major tool used in Age exploration.

Another Maintenance management framework called Total Productive Maintenance (TPM) has been developing that addresses some of the RCM age exploration intent. This is based on trying to eliminate defects and common causes of failures to improve operational effectiveness, by empowering asset operators and local maintainers to take responsibility for implementing incremental improvements that improve quality, reliability, availability, and lower cost. TPM is also linked to Operational Excellence, which also seeks to improve safety and reliability.

Key Take-Away Points

1. RCM and MSG3 share the same ancestry derived from the Nowlan and Heap RCM report of 1978. They share the same principles differing only in detail.
2. RCM and PdM are often compared as one being better than the other: This is not correct. RCM and MSG3 are a framework used to define a maintenance regime that may include PdM if the FMEA contains the right data and use of sensor time series data is considered as a means of applying on-condition data.
3. RCM has been adopted in other safety-critical industries besides aerospace, such as nuclear and the military and deserves recognition, as a strategically important system used to design optimal maintenance regimes.
4. RCM has a reputation of being overly complex and onerous. This is not deserved when RCM is run by experienced professional people. It does take effort and commitment, but the results are often transformative.
5. FMEA data is more suited to be stored in a graph database because it is rich in relationships between the data entities and attributes.

References

1. Nowlan, F.S. and Heap, H.F., "Reliability-Centered Maintenance," Report Number AD-A066579, United States Department of Defense, 1978. Archived from the original (PDF) on 1 August 2013.
2. Moubray, J., *Reliability Centred Maintenance*, 2nd ed. (Butterworth Heinemann, 1999).
3. Basson, M., *RCM 3: Risk-based Reliability Centred Maintenance* (Industrial Press, 2018).
4. SAE JA1011, "Evaluation Criteria for Reliability-Centered Maintenance (RCM) Processes," Society of Automotive Engineers (SAE), August 1, 1998.
5. SAE JA1012, "A Guide to the Reliability-Centered Maintenance (RCM) Standard," Society of Automotive Engineers (SAE), January 1, 2002.
6. Wikipedia Contributers, comparison of relational database management systems, https://en.wikipedia.org/wiki/Comparison_of_relational_database_management_systems, accessed January 2019.
7. Wikipedia contributors, Triplestores. https://en.wikipedia.org/wiki/Triplestore, accessed September 2019.
8. Wikipedia contributors, Linked data, https://en.wikipedia.org/wiki/Linked_data, accessed July 2019.

CHAPTER 7

What Are the Key Features in a PdM Maturity Model?

This chapter identifies and explains key features of a PdM system and expands why they are important. Some features that are often cited as being advantageous in marketing material (such as real time) will be critiqued and placed in a more realistic context. The chapter produces a PdM maturity model as a basis for a PdM specification or a means of scoring a third-party PdM system.

The chapter initially sets a context with an entry on the relationships between data information and knowledge, and the ambiguity that exists. A PdM system is an information system, and exploring these relationships and setting some definitions helps the reader better appreciate a successful system.

Data, Information and Knowledge

The raw material of PdM is data and, in order to understand the working of a successful PdM system, it is necessary to discuss the nature of data and how this is synthesized into information and knowledge. PdM is massively data centric and understanding data information and knowledge is enormously helpful in designing, procuring, or running a successful PdM system. A long-accepted model for data-to-knowledge was first produced in 1974, according to the reference [1], although the concepts go back further. A deeper explanation of Data-Information-Knowledge-Wisdom (DIKW) can be found at [2]. The original model was hierarchal with data on the bottom and information, knowledge, understanding, and wisdom towards the top. Each level above is a synthesis of the lower level. The original model had understanding but most literature since has omitted this as a distinct transform (Figure 7.1).

The search to find clean, simple definitions of data, information, and knowledge is hard and has not yet reached a satisfactory conclusion in literature. Knowledge is a

FIGURE 7.1 The Data-information-knowledge-wisdom (DIKW) hierarchy (Source Wikipedia) [2]).

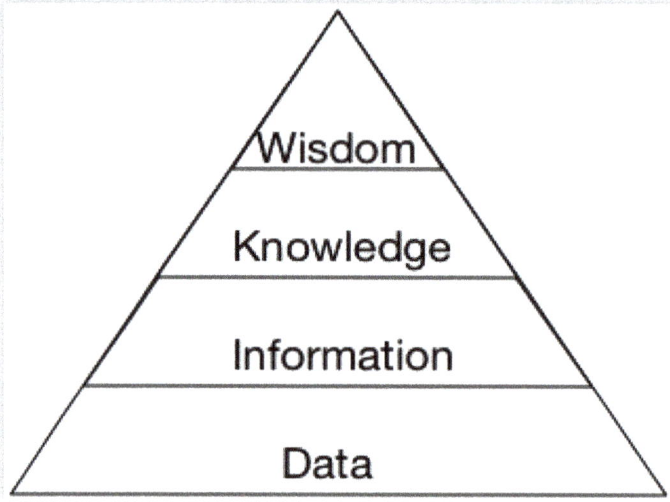

particularly elusive concept. Attempts to define it have been going on since the time of Greek philosophers, and academia still does not have a good definition that is universally accepted today. A major problem with the concepts is that the terms, such as data information and knowledge, are often used interchangeably and some attempted definitions of knowledge are described in terms of information and information in terms of data. This is wholly unsatisfactory but may never be resolved with crisp and unambiguous clarity.

There is a widespread belief that machines can infer or synthesize information from data. This is a mistaken belief as machines can give the impression that they can synthesize information. The outputs of Machine Learning and their ability to classify and regress data are very impressive, but those outputs are still not information. It is an innate ability of human beings to "sense-make" using all of their senses (not just digital data) to take in data and then synthesize information. Machines are not capable of sense making, they can only act or output new data following encoded rules. Sense-making is highly contextual, situational, and pre-disposed from previous experience of the persons along with their values and beliefs. Meanings derived from data can be entirely different for different people who are in the same place and context. Machines are unable to synthesize information.

A machine can only obey instructions, select data, infer new data, or add new labels to it. They can apply inferred data to other codified instructions and initiate actions (as described in the linked data example in Chapter 6), but they cannot synthesize information. Machines would have to be sentient to do so.

The current model to show the relationship between Data, Information, Knowledge, and Wisdom (DIKW) therefore seems to be deficient and inadequate. However, the existing model does not satisfactorily recognize the ability of machines to process and infer new data.

A solution to this was developed by Checkland and Howell [3] where the concept of Capta was introduced. Capta is a subset of data that has been selected with a context in mind, by either human or machine using encoded instructions. Capta is the missing step between Data and information that properly recognizes the ability of computing machines to process data. If this new model with Capta is adopted, it may save misconceptions that machines can synthesize information.

The other problem with then traditional model is that it includes Wisdom, which is synthesized after knowledge. Wisdom being the optimal use of knowledge to achieve goals. The author does not agree that wisdom is a true synthesis or transformation from knowledge for two reasons:

1. Wisdom seems to be a "value judgement" on Knowledge. If wisdom is merely a value judgement, it is not a true transform from knowledge.
2. Wisdom is only wisdom from certain perspectives. One person's wisdom is often another person's folly, or seeming wisdom today is recognized as folly in the future. Wisdom is therefore transient and is not ubiquitous, it is not a true transform.

In some literature, some authors make claims that Wisdom is also proof that a human has a soul. The author rejects this quasi-religious view.

The Data—Capta—Information—Knowledge model posited in this book is summarized in Figure 7.2.

If this model is used, it is possible to put the analytical and processing elements of a PdM system in their proper place from the perspective of an Information System. An Information System primarily serves human beings. Human beings are the most important component that need to be in the decision loop where stakes are high and uncertainty and variance exists. The role of computers in processing data and enabling the presentation of Capta, provides a more satisfactory positioning for the capabilities of machines and human beings.

FIGURE 7.2 The Data-Capta-Information-Knowledge synthesis model.

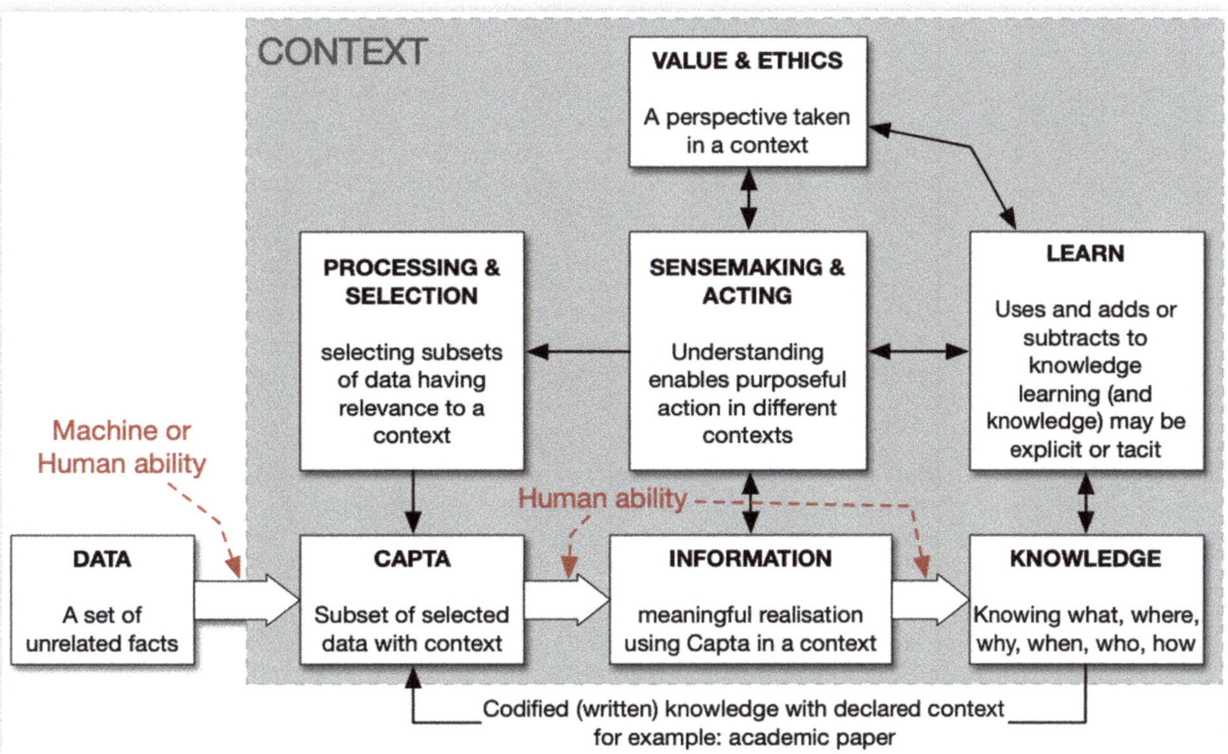

Data Quality

Data quality, according to [4], may be defined as, "*The perception of a user that data is fit for their purposes, in particular contexts.*" The important word here is perception. Quality is perceived by people and has a subjective aspect. Data may have objective measures of quality (that are essential) including accuracy, completeness, timeliness, and relevance. But if data has all of these and a user does not trust, understand, or value the data, it is still low quality.

In the PdM system, highly objective quality data and Capta must be presented to the user in a way that they can understand and act on, and the experience of using the system presenting the data must have a positive emotional impact. This experience needs to be a pleasure for them to use to be successful and complete the subjective needs of data quality.

Another vital facet of the PdM information system is that humans can take different meanings from the same data. This variation in sense-making may be minimized by the application of formal processes and shared semantics among the domain experts using the PdM system. At the other end of this spectrum of formal processes, tight teamwork carries the danger of developing groupthink, so the culture needs to be right so that people who think differently can feel confident to challenge accepted beliefs, backed up by data-driven evidence.

So what? What has this to do with a PdM system? Having a deeper understanding about the nature of data to wisdom and the role of humans and machines and knowing that data quality has both hard explicit as well as soft implicit attributes provides the reader with insights to acquire and use a PdM system.

The Breakdown of PdM into Functional Blocks

The SATAAL (Sense—Acquire—Transfer—Analyze—Act—Learn) system is a functional decomposition of the PdM system often cited with the Integrated Vehicle Health Management (IVHM) perspective on PdM. SATAAL is an acronym providing an easy to remember mental model for IVHM and PdM systems. This functional model is complementary to the OSA-CBM model discussed in Figure 3.8. Both models are widely used.

In IVHM the "Learn" part of SATAAL is not generally included. The author includes it in this book as it adds a self-reflective and continuous improvement aspect to operating a PdM system. As more data is gathered and modelling technology has improved, it is possible to apply lessons learned in constantly updating the PdM system. Constant updating and improvement should be a goal inherent in the architecture of the PdM system itself. The improvement system must include the human actors in the system.

On occasion, there may be changes in the steps taken in the process, dependent on the requirements of the system. For example, a level of analysis or pre-processing data often occurs before transfer. As can be seen in the left-hand side of Figure 3.8, if mitigation needs to occur locally on a mobile remote asset, then the full SATAAL cycle needs to occur locally. Risk needs to be managed where it can best be mitigated within the physical topology of the PdM system. In general terms, issues with shorter P-F intervals are better dealt with locally by operational means, while those with longer term P-Fs are better dealt with remotely, especially where investigative or remedial action requires central action to pre-position resources.

Sense

Sensing incorporates the sensors and measuring devices that provide the data necessary for PdM. The sensor specifications should be such that the accuracy, precision, frequency of sampling, responsiveness (stiction and hysteresis) are fit for the PdM purpose. The sense function would normally include analogue to digital conversion and the presentation of data to the acquisition system.

In the development of PdM, the requirements for sensing have been primarily driven by safety, control, and operator indication. Sensors and instrumentation are expensive to fit and maintain, as they increase system complexity and, in an aerospace context, add weight. This means that PdM systems sometimes have had to make do with the sensors and data acquisition systems fitted for these other functions. It is only recently that system design is starting to consider PdM system requirements for enhanced monitoring and data acquisition. The problems with suitability of sensors for PdM is also a factor in the difficulty in building a business case for PdM. Notwithstanding this difficulty, it is generally accepted that 80% of cost for an engineered product is expended during its service life. This is a considerable incentive to ensure PdM can have access to the right data to help minimize through life cost.

There is a PdM perspective to be developed when considering sensing, measurement, and data acquisition. One of the major considerations is data sampling rates, which is expanded below in a subsection. Other areas are:

The functions of the sensor and data acquisition. Sensors are fitted for (In order of importance)

- Protection and prevention (Interlocks)
- Control
- Operator aid—observation, alarms, and warnings
- Maintenance (which is where PdM will be included)
- Commissioning and testing. There may be mounting points or ports that may not be permanently fitted with sensors

Positioning and orientation of sensors. There are a number of physical and operating environment constraints that dictate where sensors may be fitted. The trade-off may reduce the effectiveness of PdM as it may degrade the signal to noise ratio or attenuate the signal.

Accuracy, precision, and responsiveness. PdM may have requirements that mean better quality sensors may be needed. For transient monitoring, the sensor stiction and hysteresis may be important: the frequency range for vibration sensors may need to be extended to examine features in the data at the extremes of the spectrum of frequencies.

Data Sampling Rates

We have already covered the concept of PdM systems exploit continuous data sampled from fixed sensors. The concept of continuously sampled data needs further definition. Most sensors are analogue but have analogue to digital conversions done at the sensor, so that data may be transmitted and multiplexed on data busses. Digital data therefore has an underlying rate of generation, so that they are transmitted as discrete readings over time.

The PdM system may want to consume all of the readings over time, or only some of the readings, but whatever the case, the readings are considered continuous, as a reading can be captured on demand. This section explains what the underlying sampling rates are driven by.

As part of the PdM requirements, consideration must be given to how the symptoms of failures (anomalies in the data) may be detected to feed into the diagnostic stage. These anomalies may only be detectable in four sampling rate domains:

1. **Dynamic**. Dynamic data sampling is conducted at between 1 kHz and 100's of kHz to capture vibration and acoustic data. The data needs to be transformed from time snapshots to the frequency domain using Fast Fourier and related filtering methods in order to be able to see spectral data. The various peaks of energy, velocities, or acceleration can be interpreted to extract anomalies. The frequency of sampling the raw data is determined by Shannon-Nyquist laws [5] that state that sampling must be at least twice the rate of the underlying fundamental periodicity of the feature to be detected. This is the theoretical mathematical limit: to gain practical measurements the frequency is normally five to ten times the fundamental periodicity.
2. **Transient state**. Data is sampled at rates of 1–100 Hz to capture a transient condition. For example, to capture and trend the stiction, hysteresis and cycle times of gas turbine bleed valves, where the cycle time may be 1 second, it is necessary to sample valve position and actuator power at 100 Hz. Valve cycling is a good example of a transient change.
3. **Steady state**. The easiest state to model because the relationships between the sensor data are unvarying, have reached an equilibrium and are more predictable.
4. **Chaotic state**. This is where a system may be exhibiting multiple phases (solid, liquid, gas, plasma) at the same time. Probabilistic modelling may be relevant in this state, but this state is usually avoided by control and protection systems in real life. An example where this is important comes from steam generation in a nuclear reactor core (Pressurized Water Reactor, PWR), where it is indeterminate whether the reactor coolant is in the liquid (water) or gaseous (steam) condition, and the chaotic state of multiple phases must be avoided. Modelling (digital twins) may be useful in understanding where the boundaries of the chaotic state lie, so they may be avoided. If systems move into a chaotic state, it is likely due to an unwanted excursion possibly linked with loss of control and/or protection systems.

The following diagram shows a simple system with two sets of sensor trends transiting from a steady state, through a transient state, to a new steady state. The correlations between the sensor readings become more unpredictable in the transient state, which is why steady-state modelling and splitting out the steady-state data from the bulk time series data is conducted. The relationship between values may be linear or non-linear in the steady state, but they are much more predictable compared to the transient state.

FIGURE 7.3 Trending through the transient state.

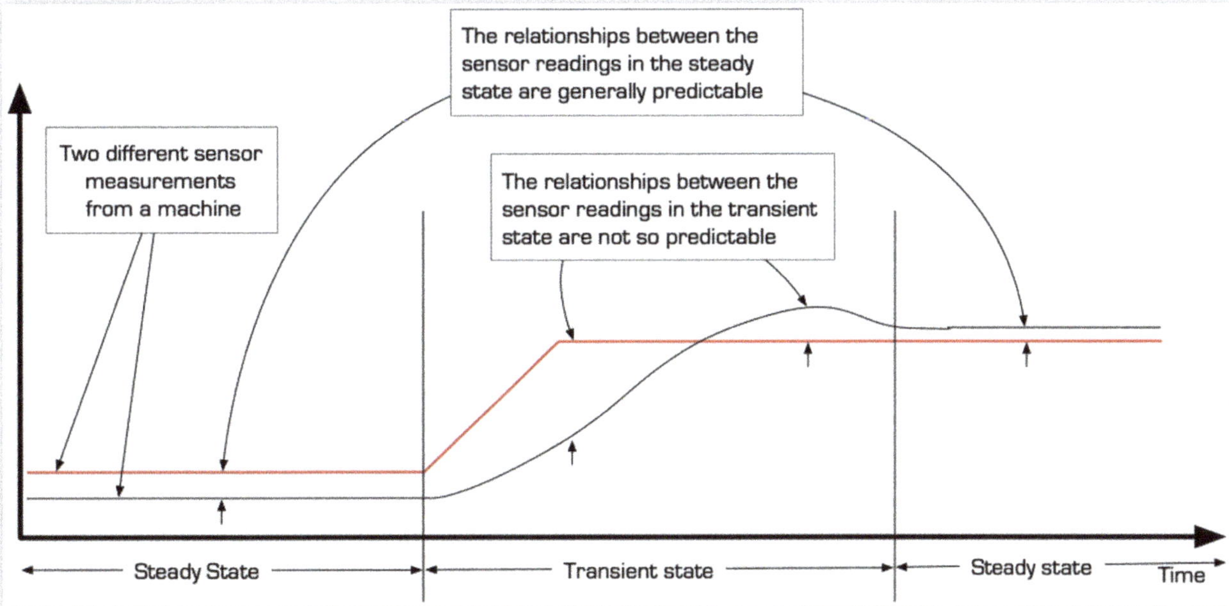

Acquire

Sensed data must be captured and made available to other functions that require it. There may be geographical, cost, and timeliness constraints that determine the physical acquisition architecture. There is often a requirement for processing data that may share attributes with the acquisition system. For example, the acquisition system may pre-process data to reduce its volume to accommodate restrictions in transmission bandwidth.

PdM systems make use of sensed data that has other primary purposes and functions. The PdM system must never reduce the effectiveness of the other primary functions that the sensors and data systems are fitted to carry out. PdM data sensing cannot be allowed to degrade safety protection or control functions.

A really important factor in acquiring time-series data is that the date and time stamps should ideally be in Coordinated Universal Time (UTC) so that there is a common baseline to compare different trends, and to avoid confusion as mobile assets cross time-zones and the international date line. For mobile assets, a geographical position should ideally be recorded: GPS can provide both position and time information.

A significant factor to be considered is time-series data reduction. Data sampled at one hertz (once a second) may be averaged and recorded for readings every minute. If machinery is in the steady state and parameters are not changing between a set of limits, then the median value is recorded whilst data stays within those limits; for example, plus or minus 1°C for a temperature reading.

If the analysis conducted is steady-state, then these sorts of data reduction regimes should be acceptable as long as the averaging times are fractions of the P-F interval. The disadvantage of doing this may be that sudden events may be masked (such as a bird-strike event in an engine). It may be desirable when averaging data to include the maximum and minimum values over the averaging period. If these values are beyond a threshold, then perhaps data may be saved at the raw sampling rate for a window of time around the event.

The data acquisition logic can be quite sophisticated in order to capture the maximum useful information. This will be traded off against the processing power, memory, and what data needs priority to transfer in a timely manner through communications channels with limited bandwidth.

Another important function is to model machine state to distinguish between steady and transient states, shown in Figure 7.3. Data sampling at the appropriate rates needs to be triggered so that the detection of anomalies can be more accurate, and the health of some components can be measured. For example, valve (transient) cycle times may indicate deterioration of stiction and hysteresis.

It is worth re-considering the discussion about operating context here. Using the example of a military front-line aircraft that has a dynamic operating profile. The aircraft are maneuvering far more frequently compared to a civil passenger aircraft flying long haul, where much of the flight sector is spent in a steady state. The PdM system for the military aircraft needs a higher percentage of transient analysis, compared with a civil aircraft with a higher proportion of static analysis. The data acquisition systems need to cater for these underlying requirements.

Transfer

The transfer of data from a mobile asset to a centralized point is necessary if timely remedial action that is better coordinated from a central position is necessary. The advantages are

- Planning is better coordinated centrally,
- The predisposition and logistics are better coordinated centrally,
- PdM, planning, and logistics expertise can be concentrated centrally and forms a critical mass that helps increase innovation of PdM system improvements,
- Whole fleet views are able to be visualized and insights are more easily noticed.

Timeliness, connectivity, and bandwidth are key constraints in the system that enables transfer of data. In aerospace, there is a worldwide network that allows VHF and satellite data communications using the ACARS system managed by SITA.

The ACARS data transfer protocol is relatively old, using a subset of ASCII characters in pre-formatted "reports" that are defined in the aircraft avionics.

Other services are starting to emerge, such as Honeywell's aircraft data gateway, that have improved interoperability, connectivity, and security.

Other ground-to-air data bandwidth is also being provided for on-aircraft on-demand entertainment, communication, and web connectivity for business users. These capabilities are being driven by improving passenger experience and expectations. Redirecting smaller volume aircraft-related data, including PdM data piggybacking on these channels, is a possibility because the data transfer infrastructure will exist.

Time-series and frequency-domain data may be compressed. Commercially (or general use open source) available compression libraries such as pkzip, or GnuZip could be used, but they are not optimized for time-series data, and better, more specific compression algorithms can be used. Other means using pictorial data sent as JPG or PNG files could be sent: for example, a selected time for a vibration waterfall diagram may be sent as a JPG, with energy peaks being encoded with color, is a viable means of highly compressing dynamic vibration data in less than 100 Kbytes. The data formats are very easy to dissect on the ground to calculate the position, volume, and peaks of the energy within.

Analyze

Analyze may be split between local processing at the asset or central processing after data has been transferred from assets. The SATAAL is a functional and not a physical abstraction.

Analysis can take two broad approaches: the first is a data science and data-centric approach, the other uses physical modelling. The Big Data approach is where all asset data is treated as one data set and analyzed as such. The idea would be to spot anomalous behavior or insights and use these to then relate back to failures. The advantage is that there are no pre-conceptions about cause and effect in this analysis. The anomalies are then normally discussed with domain experts to try to find a real cause and effect correlation.

The first stage of analyze is to conduct data quality checks, looking for missing data or obvious signs of sensor failure (a reading holds steady at a set value where there is known transient behavior or is way outside a sensible range). If time-series data is being acquired in batches, then stitching the files together to form a contiguous timeline is important. Gaps in data when the asset is out of service need to be properly accounted for. If errors or Data Quality (DQ) anomalies are spotted, then the system can alert these to the PdM operators, who may then choose downstream actions to correct them.

Many systems may convert the data into a neutral format so that the PdM system may have a single unambiguous means of analysis. This may also involve converting signals, for example converting engineering units from Imperial to SI representation.

Some PdM systems are only capable of detecting anomalies, others diagnose incipient failure. Many do not include prognostics. Some PdM systems align with Prognostic Health Monitoring, as if this is a separate system from PdM. This text posits a PdM system must include Diagnostics and Prognostics to be truly viable. Prognostics is integral to PdM.

Major stages of analysis include machine-state detection, so that data can be separated at least into transient and steady state data, Diagnosis and Prognosis. Other peripheral analysis is associated with data quality, Dynamic data is high volume and so much pre-processing and analysis needs to be undertaken on board. The time-series dynamic data needs conversion to frequency domain to understand where resonances of energy/acceleration or velocity exist. A number of frequency spectra snapshots can be taken and shown through time to derive a waterfall type view. Waterfall plots can also be separated into transient (for example rotating machinery acceleration and deceleration) or steady state (hourly readings where the data is steady). Anomaly detection can then be focused on the size and volume of predictable spikes in the frequency spectra. Other software detectors can be set up to look for unexpected spikes of energy: for example, if a machine rotates at a set speed, it will display an expected spike of energy at a predictable resonant frequency proportional to speed. If an unexpected spike of energy is observed at 0.5 and 1.5 times the resonant frequency, it is indicative (a symptom, or an anomaly) of the rotor rubbing the stator.

Part of the analysis would include normalizing the trend data using models of normality, or equipment to equipment comparisons to produce residual signals. The detection of novelties, anomalies, and symptoms can then proceed as preparation for the diagnostics stage.

Other event data may also be included, such as when restorative maintenance is conducted, that may explain why step changes may be observed in the time series data, so that any false positive diagnosis mistaking these events as failures may be actively suppressed.

Also, it may be too expensive to transfer bulk data about transient conditions in flight, so much pre-processing and analysis may be completed onboard with low-data volume summarizations and anomalies being transferred. An example may be transmitting and centrally trending timings for controlled valve cycle times. The machine state detection was mentioned in the Acquire section Act.

The PdM system can take autonomous action if it is highly trusted for some outputs. Work order and spares demands may be raised in an Enterprise Asset Management (EAM) System or Maintenance Management System (MMS). If a human must remain in the process loop, at least the work orders and spares demand may be raised in the EAM/MMS and a suitable domain user alerted to authorize their release. The mundane aspects of the work may be completed electronically.

The PdM system may alert the following classes of information. These are fundamental requirements for the PdM system.

1. Where it detects data quality issues, including missing data.
2. The performance of the algorithms, alerting when algorithms are not meeting metrics.
3. The detection of a novelty. This may be benign behavior as yet not recognized as normal behavior, the PdM operator needs to either update the model of normality or update anomaly detection if the deviation is proven not to be benign.
4. The detection of an anomaly, which is known to be indicative of non-normal behavior.
5. The detection of a symptom, which is known to be an anomaly that is a component of a diagnosis.
6. The detection of a diagnosis that indicates which component is faulty. The granularity of the fault isolation and the failure mode are the key value drivers for diagnosis. The diagnosis should also include certainty information and the probability of false positives. The use of a confusion-matrix and observer-operator characteristic curve (see Figures 3.5 and 3.6) are best practice.
7. The report of prognostic information. This may occur at any time during a long-term degradation of condition, or after a diagnosis of incipient failure. The point is to present enough information and guidance to enable operations and maintenance to make the optimal decisions based on data driven evidence. Prognostic information should ideally include:
    a. The expected, projected usage of the equipment. The usage will directly influence the rate of prognostic deterioration.
        i. It may be advantageous to show projected RUL with lighter duty cycles or usage if extensions may be operationally advantageous.
    b. The measure of condition, or resistance to failure. In the simplest case this might be a percentage of full condition.
    c. The Remaining Useful Life (RUL) left before functional failure occurs, along with certainty or confidence bands (at the time of diagnosis, the RUL equals the P-F interval).
    d. What the impact and consequences of failure will be at functional failure. There may be several points: for example, RUL to losing primary function, RUL to where induced damage to other systems is likely and the recovery work-scope grows, RUL to catastrophic failure (such as a bearing seizing).
    e. The recommended remedial action and how long the active outage time of the asset is likely to be to conduct corrective action. A medical analogy is when a doctor gives a prognosis, the patient wants to know how long it will take to recover and what they need to do before getting back to normal life. Machinery prognostics should be no different.
8. Pointers to historical corrective maintenance records for similar failure modes should be provided.

Learn

A PdM system needs constant effort to maintain quality to suppress false positives and provide timely, accurate results. There must be a feedback system from the maintenance function that conducts the troubleshooting and corrective maintenance for failures or condition reports from the PdM system. This data feeds the metrics of both the diagnostic and prognostic systems. The feedback needs to incorporate the "condition" and what failure mechanisms were involved. Rich data including image and video could be extremely useful as an archive. It is often necessary for the asset to re-enter service and new data to become available if the PdM system is able to confirm the effectiveness of diagnostics.

There is also a large opportunity for ongoing incremental improvement to the system that requires a constant effort for reflection of the results. Looking at how diagnosis can be made earlier, delivering ever better granularity of fault isolation in the components and improving the accuracy and certainty of prognostics should be part of this process as well.

The digital-twin technology, RAM discrete-event Monte-Carlo systems that have access to the full set of historical data for the sensor time-series data, along with associated maintenance records, provide a resource where active innovation and incremental improvement should be a full-time process to support the learning process.

The development of people and their skills working in a multidisciplinary team that includes data scientists, software engineers, engineering domain experts from maintenance and design are vital elements of a successful PdM system. Because of the trust issues involved with a PdM system, the way data and the auditability of the PdM system should also include User Experience (UX) experts, as the presentation of information to end users must be not just intuitive, but needs to have a positive emotional experience: a joy for a user to use. HM-1 has released AIR6915 about human factors for IVHM.

One of the key attributes of an information system such as PdM, that either recommends follow-on action (decision support), or automates follow-on action is the ability for the system stakeholders to trust in it. This is more important where the system is probabilistic in nature, where fully deterministic predictions of outcomes are not possible. Trust is subjective driven by prejudices and differing world views of the people that the system serves.

A measure of maturity for any information system is: how can the design, build, and operation of the system enhance trustworthiness? The following general principles apply:

- The analytics leading to diagnostics and prognostics should be open and intuitive to human decision makers: black-box analytics are not fit for purpose.
- The PdM system and the transforms and analytics made using it needs to be auditable throughout its data lifecycle.
- Where models are trained on data, the provenance of the models and their training data needs to be open: models, training, and verification data sets need to be version-controlled so that a history of past diagnoses and prognoses can be built up, which form a basis of continuous improvement. Regression control to move back to a previous version must also be present.

People and Competencies

In order to successfully adopt, build, and run a large PdM system, people with the following expertise and experience are required. The PdM team needs to cohesively work together, although there will be people from different backgrounds. Much work is needed to achieve a shared understanding of PdM, the system and the processes surrounding it. People from different backgrounds will have different ways of viewing and solving problems based on their own domain of experience. People need to be natural team players, who are also expert in their own areas. Brilliant people who are not team players may well slow development down. Having the whole team undertake RCM awareness training together will accelerate their melding as a team while developing a shared framework of understanding and collaboration.

The following competencies should be considered:

An Executive Sponsor: This will be a senior manager who can authorize the release of resources and who can act as champion for PdM at senior levels in the company and who will "own" the introduction of PdM. A thorough understanding of the positive business impact of applying PdM is vital to enable the executive sponsor to champion PdM. The executive sponsor will provide advice on how project reporting and business cases should be expressed to report back to senior management. Without senior management buy-in any disruptive technology adoption will have a high risk of failure.

Team Leadership and Project Management: As a leader, this person should be prepared to facilitate the team and take away their barriers. This person should also need to talk in the same language as business-orientated people and communicate technology in a simple way. This person needs to be able to manage detailed technical aspects, but also communicate with businesspeople so they understand the business reasons for adopting PdM in business terms.

Operational, Maintenance, and Reliability Expertise: Those that operate and maintain machinery are closest to machinery, and with the best understanding how they behave and fail. Senior maintainers or operators who have the ability to reflect, generalize and draw lessons from their years of practice are the ideal people to include in a PdM team as domain experts. It would also be better if these people had specialist RCM training, and could apply the principles to help implement PdM.

Product Design Expertise: While operational, maintenance and reliability expertise is more important, product design expertise is very valuable: however, designers often do not know how machinery is used operationally, and do not fully understand how machinery fares in the various operating environments and contexts that it is used in. By collaborating the PdM team can enhance the understanding of the in-service product leading to improved product design.

Systems Architecture: This would not be a full-time role and in a large organization it would possibly fall into the Enterprise architect's role. The systems architect should have experience in industrial data management and systems as well as enterprise systems. The two domains are very different. Part of the role is to extract data from the industrial systems as well as ERP, CMMS, and other maintenance-related systems so the data may be processed. Output data will need to be fed back to the transactional work-management systems so recommendations can be actioned. During the very early stages of PdM adoption where prototype and pilot systems may be set up, the architect's involvement would be guidance and learning. It is in the design and specification of a PdM production system that the architect needs to be fully engaged to ensure the system is right and properly integrated.

Software Engineering: A full stack (able to build all the layers in an application, data, logic, and presentation) software engineer with either Maths or (traditional) engineering knowledge will be ideal. Software engineers with data science, machine learning, and statistical backgrounds are highly advantageous. Software engineers should have a knowledge of software architecture.

Data Management and Systems Sustainment. Software engineers often do not understand how to support and maintain IT and data. The confidentiality, integrity, and availability of the data to authorized parties is very important. This person should also act as a mentor for the full team, as it is every persons' responsibility to maintain data integrity and quality. ISO 27000 is a well-known standard that helps define an Information Security Management System (ISMS) that ensures IT governance is appropriately applied.

Data science, statistics: Data scientists and data engineers are recently created job roles. These people are generalist in nature, having a broad background in ICT, statistics, Machine Leaning, and Artificial Intelligence (AI). For PdM there may be sub-specialization of data processing, especially in time-series data signal processing and digital filtering to improve signal-to-noise ratios. The data scientists should be able to explore what models are best suited to each diagnostic and prognostic problem and bring these to fruition in an operational PdM system. The data scientist should also be a skilled communicator, able to discuss technology with senior management.

User Experience (UX). PdM is probabilistic and needs a degree of buy-in and trust for operational and maintenance people to act on the PdM output. If diagnostic and prognostic capta is not presented in a way that is intuitive and promotes a positive emotional feeling to the users, the whole system risks being ignored. A UX expert would be massively beneficial in the success of PdM as a human-decision support system. This is an important issue to address the "Learn" section in SATAAL.

Maturity Model

The following model provides a scale for assessing the level of maturity of a PdM system:

Level 1—Data is centralized and made available for viewing and reflection. Although centralization and availability of data involves considerable investment and is a large technical achievement, it only represents the first step to PdM maturity. The volume of data for a whole asset-rich organization is usually far too much for maintenance experts to regularly review and spot anomalous machinery behavior.

Level 2—Anomaly detection. Simple alarms are set but diagnosing distinct failure modes is left to expert maintenance personnel. The kinds of anomaly detection implied by this level of maturity are described and demonstrated in Chapter 11, with the univariate Kalman Filter. The system will have a means of alerting experts who should know when........ (something is missing here).

Level 3—Diagnostics. Models for machine state and normality are incorporated and used to derive. Automated metrics for diagnostics is in-built (ideally using Confusion Matrices and RoC) (see Figures 3.5 and 3.6).

Level 4—Prognostics. The ability to predict Remaining Useful Life (RUL) are incorporated and used to estimate the times to varying degrees of functional failure. Metrics for prognostics are in-built.

Level 5—Continuous improvement, verification and auditability are built into the PdM system. Processes are embedded for the whole PdM system and for who, what, and when actions should be initiated.

Key Take-Away Points

1. Specifying a PdM system must not be considered a primarily technical issue: introducing disruptive technology always initiates change (Systems Theory and emergent behavior). This change can be anticipated and managed or ignored.
2. Knowledge, culture, and ways of thinking about probability are necessary to embrace PdM.

References

1. Wallace, D.P., *Knowledge Management: Historical and Cross-Disciplinary Themes* (Libraries Unlimited, 2007), 1–14.
2. Wikipedia Contributors, "DIKW_pyramid," https://en.wikipedia.org/wiki/DIKW_pyramid, accessed August 2019.
3. Checkland and Howell, *Information, Systems and Information Systems, Making Sense of the Field* (John Wiley & Sons Ltd, 1998).
4. Dibsdale, C.E., "Effective Tacit Knowledge Sharing between Knowledge Workers Who Identify Anomalies in Data," MSc dissertation with the Open University UK, 2010.
5. Wikipedia Contributors, "Nyquist-Shannon Sampling System," https://en.wikipedia.org/wiki/Nyquist%E2%80%93Shannon_sampling_theorem, accessed January 2019.

Specifying Predictive Maintenance

This chapter includes a template for judging the maturity of an organization that relies on complex assets for its core activities. It also looks at requirements specification where relevance or assumptions to the maturity matrix will be highlighted. This may mean that several important investments may need to be made before a strategic Predictive Maintenance (PdM) system is introduced. Even if an organization is relatively immature in terms of readiness for PdM, there may be areas where early opportunities to prototype pay back handsomely. Prototyping allows lessons can be learned to help de-risk a larger strategic investment in the roll-out of a full PdM system. Quick wins that have high value establish credibility in the eyes of the senior management and confidence in the minds of the consumer. Care must be taken with deploying prototypes, to deter business people to assume that prototypes are production capable. The limits and costs of trying to scale prototypes must be carefully and robustly explained.

Assumptions

The most critical assumptions are that the organization relies on physical assets that:

- business critical, so there is a high reliance on the assets to achieve the organizations goals.
- have high impact or consequences to the business or other stakeholders if they fail.
- require maintenance.

This implies that assets do not necessarily have to be complex, capital intensive (relatively expensive), or require specialist maintenance possibly recommended by the asset manufacturer. But assets that do have these attributes are most probably going to require a maintenance system.

Another case where maintenance needs to be considered is if assets are used in a novel operating context, or environment, that the manufacturer has not considered in their maintenance recommendations.

Designers and manufacturers survive in the market because their assets have been designed to be safe and reliable (meeting standards and being competitive). However, although the asset may be intrinsically safe, the operational safety may need wider consideration as it, depends how assets are used or integrated into a larger system. This is especially important in an industrial situation, where vulnerabilities that may have safety implications emerge (remember the principles of systems thinking and emergent behavior).

Having an FMEA to support and justify a maintenance regime is not mandatory, unless it is claimed to be an RCM process but experience suggests that it is highly beneficial in applying PdM. Especially where assets are used in novel context and environments.

Note that SAE standard JA1011 [1] specifies what is necessary in a process to claim that it is compliant. RCM mandates the use of a FMEA to design a maintenance regime.

Basic Requirements

The list here captures the requirements of PdM systems using the SATAAL breakdown and their explanations. The requirements produced here are either "Must Have", or "Should Have" where "Must Have" is critical for the system to work, and "Should Have" is a function that adds considerable value to PdM system. This system of marking requirement importance uses some of the MoSCoW system used in software engineering.

- M = Must Have. If these requirements were missing, the system would not function at its highest level. These requirements are critical.
- S = Should Have. If these requirements were missing, the system would work but be severely compromised.
- C = Could Have. These requirements are nice to have and may be included if budget and time constraints allow.
- W = Would Have. These will not be done in this phase of the project as they are only partially beneficial; but may be retained for review later.

If any "Should Have" requirements are missing, they severely degrade the PdM system. The following are purely functional requirements as performance, constraints, and other non-functional requirements will be contingent on the context of the system being considered. There may be constraints due to existing systems to interface with and ITC infrastructure that will all determine non-functional requirements. This is a set of "Must Have" and "Should Have" requirements that may be used as a baseline:

- The PdM system shall be able to connect to and process time-series, event, and frequency domain data (Must Have).
- The PdM system shall be able to apply state detection algorithms to split data into transient or steady states (Should Have).

- The PdM system shall be able to, if relevant using vibration, apply signal processing to transform frequency to time domain and vice versa (Should Have, but is more critical with rotating machinery).
- The PdM system shall provide means to explore data, to find correlations between and within event and trend data (Should Have)
- The PdM system shall be able to use models representing normal behavior (Should Have, but this technique is a significant differentiator).
    - Models may be physics- or data-based.
    - Models may be either imported from an external source or data-driven models built and trained using the PdM system itself.
    - Where data is used for models (build, training, and verification) data sets must be separated and labeled. These data sets must be version controlled and be available for historical auditing purposes.
- The PdM system shall be able to produce residual signals from normal behavior models and sampled data (as above—Should Have).
- The PdM system shall be able to detect deviation from normal behavior and classify novelties, anomalies, or symptoms. These algorithms will include threshold exceedance, ramp, and step changes in the time series trends. The anomaly detectors shall be tuneable by the users (Must Have).
- The PdM system shall be able to track anomalies through time and classify these where they form a signature of a failure mode (diagnosis) (Must Have for a complete PdM system) this classifies anomalies as symptoms of a failure mode.
- The PdM system shall be able to detect failure early enough and alert those needing to intervene in sufficient time so that they can take effective remedial action. (Must Have).
- The PdM system shall be able to isolate failure modes and failed components at the required level of indenture in machinery break down structure to be able to facilitate effective and efficient recovery. For example, should the system fault isolation be at the whole engine, compressor combustor turbine or the row of blades or guide vanes levels of indenture (Must Have).
- The PdM system shall provide a means of measuring the performance of diagnosis (usually a Confusion Matrix) (Should Have – but improving the system depends on measuring its performance).
- The PdM system shall be able to provide prognostic regression to report Remaining Useful Life with certainty bands, along with forecasts of effects and consequences of functional failure (Must Have).
- The PdM system shall be able to apply any prognostic model (Should Have).
    - Models may be physics- or data-based.
    - Models may be imported from an external source or data-driven models built and trained using the PdM system itself.
    - Where data is used for models (build, training, and verification) data sets may be separated and labeled. These data sets must be version controlled and be available for historical auditing purposes.

- The PdM system shall have an alerting system that may be user configured to inform those responsible for decision making or taking remedial actions, with settable priorities, for anomalies diagnosis and prognostic events (Must Have).
  - Alerts shall be able to be positively acknowledged to ensure they are received.
  - Actions taken on alerts shall be recorded in the PdM system.
  - The PdM system shall be able to form work orders for a maintenance scheduling system (leaving a maintenance scheduler user authorization to release the work order).
  - Make available all historical alerts of the same nature, to reassure the users that the appropriate action is initiated.
  - Ensure that alerts are not repeated for the same failure event, where no significant changes have occurred.
- The PdM system shall share data to enable Age Exploration, to establish parts age and usage in relation to condition and failure (Should Have).
- The PdM system shall be capable of advising other opportunistic maintenance when a failure is diagnosed, where the assets recovery provides the opportunity for other work that may be due, or where prognostics indicates a high probability of failure during the next phase of operations (Should Have).
- Any changes within the PdM system shall be recorded, by whom and when, along with a justification for making changes. This shall form an audit trail (Should Have).
  - Some changes may only be available to named users
  - Changes shall be able to be reversed
- The PdM system shall be able to archive older data (Should Have).
  - Archived data shall be able to be restored into the PdM system on demand
- The PdM system shall be able to produce reports for any failure, with anomaly detection, diagnostics, prognostics, and alerting events (Should Have).
- The PdM system shall be able to be used as assistance to Root Cause Analysis (Should Have).
- The PdM system shall be able to report any asset's components likelihood of failure and Remaining Useful Life to support maintenance planners searching for candidates for opportunistic maintenance (Should Have, inclusion of Type 1 and 2 prognostic capability).
- The PdM system shall be able to export data for offline analysis (Should Have).
- The PdM system shall be able to separate data logically between different stakeholders so preserving privacy (Must Have, as this may also have legal ramifications if personal data is included).
- The PdM system shall include a feedback system so the maintenance function conducting corrective maintenance can report on the condition of the part to verify diagnosis and prognosis (Should Have, very important for verification and continuous improvement).
- The PdM system shall provide case studies and be able to be used for training users (Should Have, but very important users are considered as intrinsic to a successful PdM system).

- The PdM system shall interface with data historians, system surveillance (for example SCADA), reliability applications, FMEA structures, FRACAS (Failure Recording Analysis and Corrective Action Systems), maintenance management and logistics systems. (Should Have—very important for automation, transfer of data involving manual input is not scalable).

Key Take-Away Points

1. This chapter provides organizational context for needing a PdM system.
2. It provides baseline set of functional requirements for an effective PdM system.

Reference

1. SAE JA1011, "Evaluation Criteria for Reliability-Centered Maintenance (RCM) Processes," Society of Automotive Engineers (SAE), 1 August 1998.

What Are the Disadvantages of PdM and How Should They Be Addressed?

PdM is firmly based on and subject to probability. Many other processes and mindsets are based on determinism and deterministic outcomes. People who have spent a major part of their careers tacitly learning how to deal with their world in a deterministic manner face a substantial challenge here. They have to adapt their thinking and reflect on and change long-held assumptions when dealing with the probabilistic nature of PdM. This change is necessary, and it demands change at every level of the organization: this is not an easy cultural move. This problem is especially acute in building business cases for PdM.

There is useful literature to be reviewed in understanding how people think, which provides knowledge for how culture change may be planned so that organizations can adapt to dealing with probability.

Daniel Kahneman: Thinking Fast, Thinking Slow

Khaneman book [1] describes how the brain is organized into two working systems. System 1 is the fast-acting intuitive side that works subconsciously and underpins how people automatically react to danger without conscious effort. System 2 is rational, conscious, and slower acting. System 2 takes more energy and effort to use, and thus our brain combines System 1 and System 2, preferentially using System 1 to make quick choices. This natural tendency may not be the best way of making decisions where a more considered and rational approach is necessary. The book describes how the brain also compensates for not being able to process and store all of our sensory inputs. It also shows how decision-making can be highly influenced by external stimuli. Reading this and understanding how our mind works work may help make us better decision makers. Our brain is limited because it does not remember every single past event: the brain actually superimposes frequently observed patterns it has seen in the past to current situations, which means people may not notice other peripheral things going on. An example of this "selective attention test" is found at [2]. People can easily be biased in misinterpreting data or capta because of erroneous patterns recalled from our memory.

Nassim Nicholas Taleb: Fooled by Randomness

This book [3] has stock market investment as its context, where investment success and decision making are far more influenced by random chance than most of us give credit for. There are widespread beliefs in extreme success being due to expertise where in reality it is random luck. The book contains some strategies that are useful in dealing with probability and expectancy. One of these directly relevant to PdM decision making is where a maintenance and operational machinery manager has a diagnostic alert with prognosis advice from the PdM system. The dilemma is that whatever decision is made whether to act or not, the costs are high:

- The cost of withdrawing an asset from service and conducting the corrective maintenance when the diagnosis is false may be unacceptable.
- The cost of ignoring a diagnosis and suffering an unplanned failure defeats the object of applying PdM. This cost is unacceptable.

The decision making is critical and has high intrinsic value; understanding more about how humans make decisions is valuable to enable the decision maker to have multiple perspectives to help them decide. The book illustrates how decision making may be based on risk management principles to help insulate yourself against dire outcomes.

This is also relevant in some RCM processes where an FMEA Probabilistic Risk Number (PRN) criticality matrix is used (discussed in Chapter 5) to determine the criticality of failure based on the likelihood and impact of failure. If the RCM process is considered where expert knowledge is being elicited from experienced maintainers, then the process may insulate the organization from failure events that have dire consequences, but these events may be so unlikely that some may consider them as not credible. If an event like this occurred, it may be considered a "Black Swan" event [4]. The mitigation against extremely high impact/extremely low likelihood events is to relax the likelihood parameter (because likelihood is much harder to accurately estimate compared with impact), and apply active measures such as employing PdM specific to that failure.

The point is that management and decision making in the probabilistic domain require a change of mindset, and it is highly advisable to build some background knowledge about how human decision making is made to deal with uncertainty. Probability and uncertainty pervade PdM.

The certainty of predicting future needs to withdraw assets from service for maintenance reduces with PdM. In a world where strict periodicities are set for planned maintenance (scheduled inspections, replacements, and restorative tasks), the forward-looking plan has relatively high certainty. This means capacity planning, utilization of maintenance resources, financial commitments, and minimization of disruption to asset operational schedules can be forecasted with high confidence.

A clear plan based on fixed period scheduling looks very neat and outwardly attractive, but in common with any military operation, a plan is only valid up to the point where battle is joined. Once a set of assets starts to have unplanned failures, the neatness of the original plan quickly disappears; and the planning process can become an ongoing fire-fighting for the recovery from unexpected failure events.

With PdM the prediction of failure has uncertainty, and this is also inherited into any plan. The plan looks messy, but in reality, it should (if PdM is effective) actually represent the real-world demands for maintenance intervention compared with a fixed-period schedule. The key to success is where PdM delivers sufficient P-F interval or Remaining Useful Life (RUL) to allow planning and predisposition of resources for recovery.

The possibility of false positives causing nugatory work may lead to a loss of confidence and trust in PdM. The continuous effort to improve PdM and suppress false positives is considerable. Loss of trust in PdM can occur quickly with a small number of false-positive events involving unnecessary withdrawal of assets from service and useless work. This may take a long time to re-establish. A sensitive trade-off needs to be continuously reviewed against balancing specificity and sensitivity of the diagnosis process. On occasion, it may take manual confirmation of the diagnosis (by inspection) to ensure false positives are avoided but this checking activity tends to reduce the cost effectiveness. The use of a Confusion Matrix and Receiver Operator Characteristic curve (ROC) to measure and trade-off diagnostics is highly recommended.

Maintenance planning and scheduling may be perceived to be less predictable than calendar of elapsed operating hour or cycles-based maintenance. This makes planning and scheduling maintenance harder, along with problems of utilizing maintenance resources in a steady manner more problematic. Spares forecasting and ordering may require that larger stocks are held and slightly increased costs with associated sunk capital.

A mature PdM mitigates some of these disadvantages by focusing on delivering a reasonable diagnosis of failure early enough that the Remaining Useful Life (RUL) is maximized, and optimal times where assets are withdrawn from service may be made to minimizing disruption. Most maintenance regimes have concepts of planning horizons as follows:

1. Short-term planning horizon, possibly two weeks to a month looking forward: maintenance plans are solidly committed, and spare parts are predisposed (or firmly committed) in order to conduct work. If an organization mandates that RUL exceeds the short-term planning horizon for any failure mode, then the disadvantages of more uncertain planning and scheduling are mainly mitigated.
2. Medium-term planning horizon, possibly three to six months in advance (or two minor maintenance cycles ahead, where machinery is regularly withdrawn from service). This is where maintenance is planned but not yet committed.
3. Long-term planning horizon, possibly looking at a year ahead, and likely to be linked to the annual financial planning and commitment cycle. This is where longer term capacity planning is likely to take place having understood the utilization of maintenance resources and how their budget is consumed.

The willingness to continue operations with known incipient failures is low (in aerospace). Any accident caused by a known incipient failure will cause catastrophic loss of reputation and loss of confidence in the whole aerospace industry. In this situation, the ability to continue operations with a known higher risk of failure is probably untenable.

If a serious defect has been found with a low probability of functional failure for several flights and the aircraft has no local maintenance facilities, it may be justifiable to conduct a ferry flight with minimal crew and no passengers to an airport with the appropriate maintenance capability. If incipient failure is going to be corrected as soon as possible after diagnosis, then most efforts will be put into the accuracy of diagnostics, with prognostics being of lesser importance. (This does not imply prognostics is not important.)

In other industries, where production is a priority and there are no safety implications, machinery with known damage may be kept running for as long as possible. In extreme cases, a machine may be run to failure, because from a business perspective it may be more beneficial to do so. Perhaps there are large penalties for not meeting a production dispatch commitment. Having effective and trustable prognostics is vital in these situations.

In the industry a reader works in, the tolerance to continuing operating with known defects must be assessed and extra effort must be put into the continual improvement of prognostic accuracy.

Access, ownership, and security of data. Many organizations are realizing the inherent value of data and its utility as a business resource. The questions of access and ownership of data derived from assets is becoming sharper. In some industries manufacturers sell access to in-service data to operators. This practice will likely become untenable, as that data can easily be analyzed and the whole picture of how assets are utilized and operated can be derived, which constitutes sensitive Intellectual Property (IP).

If whole industry verticals and the data involved in the full life cycles of complex machinery are reviewed, it will be noticed that data exists in siloes. Many data silos will belong to different stakeholders within the industry. Design and manufacturing, operators (including PdM) and R&O data are most often separated and not shared. Systems to exploit the data by its aggregation have not yet been produced, and the structure of the data held by each organization is different and not interoperable.

If a means by which data could be shared technically, so it is interoperable and transferrable, that takes into account sensitivities that protect IP, it would have the potential to save much through-life cost.

There is a difference in structured data and unstructured data. Structured data may be contained in relational database systems while unstructured data (images, videos, and documents) are held in hard copy or in computer file systems. It is estimated that only 5% of all data held by an organization is kept as structured data in a database management system. Data schemas to contain data suitable for industry are often not stored in a standard way, although data standards are developing. As Big Data technology has been emerging there are new types of what are called "no SQL" database management systems. No SQL systems fall into five basic types:

- Key Pair data stores.
- Document databases.
- Time-series databases. This would include industrial historian systems that store sensor time-series data.
- Big table (column or row orientated) databases.
- Graph databases. An application for graph databases is discussed in Chapter 6.

It is not in the scope of this book to provide a detailed explanation of these new systems. It is worth knowing about them because it does imply that a much higher percentage of all data held by organizations may be stored more securely in No-SQL databases. This makes that data available to be exploited as a valuable business resource.

Security of data is going to require constant effort and work to ensure only those authorized to see or access it can. With the emergence of Industrial Internet of Things (IIoT) and ever greater networked connectivity of machinery and components the vulnerabilities associated with a network multiply.

The ability for PdM that relies on connectivity, to deliver maintenance credits must include the highest standards of data security for it to be successfully deployed. The current standards applied to current data systems are not yet fit for purpose. SAE has a Digital and Data Steering group (DDSG) to analyze and report on gaps in existing standards. Certifying a PdM system to ISO 27000 is also a means of addressing security.

The adoption and full exploitation of PdM requires fundamental changes. Any change in an organization or federation of organizations and stakeholders is uncomfortable and difficult. The amount of effort required to invest in the change management process is large and may easily be underestimated.

It may be tempting to cut project costs by reducing or eliminating change management and training, which actually increases the risk that the necessary deep-rooted changes are not made, and PdM does not deliver value. Usually, social and human aspects of a technology introduction are harder to get right than deploying the technology itself. If a project is initiated and leadership is devolved for delivering it to IT or engineering departments, then the likelihood that the project will fail is high. Successfully adopting PdM must be business lead and is mainly a people-change issue, the technology aspect often being the easiest to build.

The Resnikoff Conundrum

In his book RCMII, Moubray [7], described the conundrum that the higher the impact of failure (especially safety-implicated) the rarer will be the failure events. It is also not ethical and most often too expensive to test to failure where safety impacts are apparent. This means there may not be any data of statistical significance about these events. Moubray concluded that actuarial analysis was therefore not practical in these cases and where data was required to understand the failure modes, he recommended it be derived from expert operators and maintainers through expert elicitation with a skilled RCM facilitator.

Whilst knowledge elicitation from experienced operators and maintainers is vital and should be the default position for an RCM study, the introduction of modern information systems, metrology and communication technology coupled with the availability of data, have improved at an exponential rate over the last 10–15 years. The author believes Moubray's conclusions may be challenged for the following reasons:

- The application of Bayes' Theorem and its recent ascendency from being the underdog of the frequentists school in the statistical community has been transformative. The ability to model physical processes and degradation is way beyond the levels assumed by Moubray in the 1980s.
- Besides Bayes, the emergence of Artificial Intelligence and Machine Learning that underpin data-driven predictive technologies provide the tools to model and predict machinery behavior.
- The emergence of digital twins utilizing detailed physical and data-driven models enables flying a physical and virtual asset synchronously: any abnormal behavior may be explored using the digital twin, enhancing root-cause analysis, diagnostics, and prognostics.

- Moore's Law with the capability of IT doubling for the same cost every 2 years since Moubray's time, the emergence of cloud, distributed computing-on-demand using Big Data technology have transformed the ability for computing to process and solve huge problems. Consider how weather forecasting has improved over the last 20 years by this development, where five-day forecasts may now be trusted. Strictly speaking, Moore's law that predicts the increase in the density of transistors per unit area of silicon has reached physical limits. The processing capability is still increasing, because multiple processors are being produced per Central Processing Unit (CPU), and programming languages can exploit this using parallel processing to reduce the time taken to compute.
- We are also at the point where the Industrial Internet of Things (IIoT) is emerging, where miniaturized, smart, ultra-low power, wireless-enabled sensors and devices will slash the costs of fitting sensors by orders of magnitude compared to today's systems. The result is likely to be far more reliable with data becoming available to monitor and measure, or accurately infer condition degradation. IIoT is an emerging technology, where barriers for adoption will be discussed in Chapter 10 but the potential is transformative.
- The following may be accessed and analyzed: used parts that may be deeply inspected or run to failure; accessing data captured during their service life to determine the degradation of condition including advanced prognostics. These will supply detailed capta to predict failure. These data combined with Bayesian, digital twin, and other machine learning (applied deep learning perhaps) may address Moubray's criticism.

The conclusion is that by exploiting more of the data that is created through the lifecycle of the components, it should be possible to conduct more actuarial-type analysis to specify and help justify the configuration of a PdM system to increase the confidence in its outputs.

However, access to the data may still be problematic as data may be owned by different organizations in the industry. Manufacturers and their suppliers, operators, maintainers, and other parties may be reticent to openly share data as this may expose IP. Organizations are also becoming more aware of the latent business value in their data and want to ensure that they gain the benefits of analyzing it.

Commercial incentivization to share data where benefits accrue to all organizations needs to be worked out. The technology to enable sharing is available: it is the willingness to share data between organizations within industrial sectors, whilst preserving Intellectual Property and value streams, that needs to be developed.

Key Take-Away Points

1. There are disadvantages in PdM:
    a. The need to deal with probabilistic systems that requires adjustment in thinking and decision making.
    b. The changes in projecting future maintenance are made uncertain by PdM, but this is mitigated by having enough time within the Remaining Useful Life to predispose resources and plan for recovery.
    c. New competencies are necessary to employ and develop.
2. Modern ICT and data analytics, along with increases of data availability, start to challenge the RCM belief that actuarial analysis is not practical.

3. Commercial issues in data ownership and access need to be overcome in adopting PdM, without compromising sensitive IP.
4. IVHM has a large potential to reduce costs throughout the supply chain, and also in the product lifecycle.

References

1. Kahneman, D., *Thinking, Fast and Slow* (New York: Farrar, Straus and Giroux, 2011).
2. Chabros, C. and Simons, D., "Selective Attention Test," http://www.theinvisiblegorilla.com/gorilla_experiment.html.
3. Taleb, N., *Fooled by Randomness: The Hidden Role of Chance in the Markets and in Life* (New York: Texere, 2001).
4. Wikipedia Contributors, "Black Swan Theory," https://en.wikipedia.org/wiki/Black_swan_theory, accessed August 2019.
5. Wikipedia Contributors, "Confusion Matrix," https://en.wikipedia.org/wiki/Confusion_matrix, accessed July 2019.
6. Wikipedia Contributors, "RoC Curve," https://en.wikipedia.org/wiki/Receiver_operating_characteristic, accessed July 2019.
7. Moubray, J., *Reliability Centred Maintenance*, 2nd ed. (Butterworth Heinemann, 1999).

CHAPTER 10

How PdM Will Likely Transform with the Emergence of New Technology

We live in an age where the emergence of new technologies and their universal and rapid adoption is accelerating (see Figure 1.1). An example of this is the invention and take-up of the landline telephone system that took just under 100 years before it was fully adopted in developed countries. This should be compared with the mobile smart phone that has emerged and been universally adopted in the last 15 years in most countries in the world.

Big Data and Cloud Services

The first technology discussed that has recently emerged is Big Data and the availability of internet-based cloud services.

Big Data technology has been developed because of the requirements of on-line (internet-based) commerce and information services. Amazon.com required reliable online ordering and fulfilment systems that also directly marketed to customers based on analysis of millions of users' buying preferences. Google required search-based services to return personalized and useful information to users in under a second. The requirements for much higher volumes and varieties of data, to be processed at much higher speeds (velocity) and timeliness, where the outputs were accurate and could be trusted (addressing veracity) were necessary in order to deliver usable trusted services. See Ref. [1] for the IBM explanations of the four "V"s in Big Data. The four Vs are volume, variety, velocity, and veracity.

The traditional data management vendors did not have systems or products available to economically deliver the performance demanded with volume, variety, veracity, and velocity from the new online businesses. These businesses developed the Big Data technology themselves but used an open-source approach. The technologies they invented helped scale computing infrastructure, producing distributed and parallel processing. Core technologies like:

- Hadoop [2] that enables storage and processing of vast volumes of data includes a file system,
- Hadoop File System (HDFS), that allows data to be split over very many cheap file servers using a big table No-SQL approach (discussed in Chapter 6, using a No-SQL graph database).
- The development of Map-Reduce algorithms [3] added the distributed parallel processing that substantially reduces the time to compute over massive distributed data sets on Hadoop.

The traditional data management solutions had to include very expensive server clusters, which grow exponentially in cost for marginal improvement at the upper end of the spectrum of capability.

There is much hype about industry adopting Big Data, cloud services, and IoT: many projects including PdM are being launched claiming exploitation of Big Data. While these claims may be exaggerated and much of the projects based on current technology (such as conventional sensors), it is still legitimate to use the new emerging information and communication technology (ICT) capabilities: Organizations do not necessarily need Big Data to exploit Big Data technologies. Many projects experiment with Big Data to position themselves for downstream exploitation and allow experience to be gained.

Compared with the online giants such as Google and Amazon, most aerospace companies are not in the same league in terms of Big Data, but this does not stop aerospace from taking advantage of and exploiting the technology.

The Emergence of the Industrial Internet of Things (IIoT)

IoT and its industrial equivalent means that any smart device with localized processing and memory may be connected to the Internet [4]. Smart devices may also locally network with other IIoT devices using ultra low-powered mesh wireless networks. These local networks (in IoT terms called the "edge") can act autonomously and adapt depending on what they sense from their environment. The edge-networked devices will be able to do more together than they could do as the sum of their isolated abilities. For example, in an IoT-enabled in-house motion sensors may work to switch room lights on and off as people move through rooms. The system will also distinguish between authorized and unauthorized people, so the system does not inadvertently help burglars.

In the industrial sector, IoT needs more maturity and have a market developed for volume sales in the following areas:

1. Security needs to be seriously upgraded. There is an adage "*if it has software it might be hacked, if it works on a network it will be hacked*".
2. Industrial packaging of the sensors is necessary to operate in harsher environments in industry.
3. The trade-offs between power, processing, memory, and data transfer (bandwidth, power, and range) need optimizing for industrial purposes. Most systems today will be bespoke, and commoditization will be necessary before a large uptake ensues.

4. The early adopters of the technology will more likely be those organizations with static assets. For example, oil and gas exploration are prime for this as instrumentation associated with drilling may provide early breakthroughs with hardened packaging.
5. Predictive analytics will need further development so that the algorithms will work with reduced processing power and memory. It might emerge that specialized low-power processing chips (such as Google's Tensor Processing Units, TPU) may be used in IIoT edge applications.

The promise of IIoT is that the cost of fitting sensors reduces by orders of magnitude compared with today's technology. This provides the opportunity to measure, sense, and make data available at a far greater granularity than is achievable today. Sensors will be miniaturized adding a step improvement in the ability to conduct Predictive Maintenance. Miniaturization is possible using micro-electrical mechanical sensor (MEMS) methods and minimally invasive monitoring sensors (MIMS) being developed in the medical domain, where the principles can be adapted for use on machinery.

Initial applications with IIoT in aerospace may emerge in structural and avionic-type systems. It will probably not emerge in gas-path sensing in jet engines due to extreme environments.

IIoT will also enable better PdM architectures, conducting more onboard processing to deliver more local recommendations and automatic actions based on the ability for local processing to mitigate incipient failure compared with centralized functions. Local automation of diagnostics, prognostics, and booking remedial action and initiating planning and logistics systems to meet the maintenance demands will become common. If self-driving vehicles become more common, then the advent of them turning up at maintenance bays with remedial work already scheduled may become prevalent. The automation potential of IIoT still needs to be balanced against the requirement of centralized PdM systems components. The human needs in an information system are still paramount, and the concentration of expertise in a critical mass enables robustness learning and continuous improvement.

IIoT will also be optimally positioned to take advantage of 5G mobile-phone technology [5]. 5G enables local wireless networks of smart IoT devices to communicate large data sets with reduced latency (compared with 4G systems). IIoT may be the killer application of 5G telecommunications systems with the maturity of 5G rollout, the author may begin to believe marketing hype about real time PdM.

Industry 4.0

Industry 4.0 [6] is an amalgamation of several emerging technologies including IIoT, robotics, Big Data, additive manufacturing, data-driven manufacturing, and predictive technologies. It promises to deliver highly autonomous production lines that may reconfigure themselves (using data and instructions) for manufacturing different parts. The system can respond to short-term demand and be better suited to Just in Time (JIT). Production line workers are minimized, but new skilled digital jobs to support the new systems will develop.

As part of the IIoT the smart production line may also electronically negotiate supply of materials and use centralized optimization intelligence to produce the most economic production runs, optimizing switching time, and costs between reconfigurations.

This technology may also impact globalization because the wages of production line workers is less important and the emphasis is on creating software (to both define the manufacturing instructions for the product being produced, and for the configuration of the production line itself). Manufacturing may become more localized, producing goods where they are demanded, thereby reducing transport costs and delays. It may also reduce risks of stock-outs, perhaps allowing contingency stocks to be reduced as well if they can be locally manufactured just in time on demand. The transformative change that Industry 4.0 promises is that data becomes the export commodity instead of goods. This changes the nature of value chains and allows costs to be reduced.

PdM should be an integral technology to Industry 4.0, with a prime task of monitoring and assuring the quality of the manufactured product. If product quality is the primary target for PdM in an Industry 4.0 production line, then the health management of the manufacturing equipment will be delivered as a by-product.

The potential for aerospace manufacturing is attractive, but it is the higher potential to share data that also has benefits. Data produced in the Industry 4.0 process should be available to product in-service PdM. Access may provide datum points for establishing condition. Access to design and manufacturing records may provide knowledge of concessions to design that may affect intrinsic reliability. Industry 4.0 data is also useful in enriching digital twins (providing test data, etc. from products).

Nanotechnology

The application of nanotechnology to structures and materials will be continuously developed: new materials will have different behaviors to failure mechanisms (corrosion, fatigue, etc.) that will require new methods for inspection, testing, and measurement of condition.

One exciting opportunity is the ability to weave glass fiber into structures where the fiber has inherent temperature and strain measuring capability. Many sensors may exist in series in a single strand of fiber. The ability to measure temperature and dynamic stresses in a matrix of sensors embedded inside structures presents exciting opportunities to eliminate invasive inspections or NDT (Non-Destructive Examination or Testing). This technology has the potential to augment and eventually replace aerospace ABC type checks [7] where extensive invasive manual inspections could be replaced with continuous PdM monitoring. This has huge benefits in reducing the high costs of these checks and increasing aircraft availability.

A recently emerging technology is in the field of miniaturized robotics where inspection and image capture of inaccessible places become possible. The technology may also be developed with additive manufacturing to deposit new material or dress and improve worn surfaces to restore their condition. The data gathered from such devices augment the condition information available to Predictive Maintenance.

Configuration Management

One of the underlying technical enablers of the IoT is the adoption of the IP6v [8] addressing space for indexing the internet. The older standard IP4v, allowed for over 4 billion unique addresses (2^{32}) to be indexed in the internet, and recently this capacity has been used. IPv6 has 2^{128} addresses available. Not all addresses are available for general use, some being reserved, but IPv6 is essential as an IoT enabler to allow billions and billions of new smart devices to be directly connected to the internet.

If a review of Internet Protocols (IP) and internet addressing is undertaken, extending to the world-wide-web then the URLs and URIs (Uniform Resource Locators or Identifiers) are translations from the IP address (expressed as groups of numbers) are defined by humans and are consequently much more human friendly to read.

The beauty of using IP addresses (URI & URL) is that:

- all parts may be assigned a unique IP address. It is possible to use IP masking to differentiate between the equivalent of a part number and a part serial number.
- Each part is uniquely addressable and communicable over the internet, if it is connected with any associated sensors or RFID tag. The part may be traced.
- Each part may use extensions to the URI and URL to point to any data about that part (again accessible via the internet). The part URL not only identifies the part-serial number, location, current data, but also any associated historical data. The part carries its own log history with maintenance and certificates of conformance.
- Using linked data or data triples, a network of relationships can be modeled that provides rules for any type of relationship between individuals. Physical or functional dependencies may be modeled and maintained. This enables complex rules for part effectivity to be maintained. Many assets, systems, and components may be improved and modified. There are complex rules associated with configuration management. They ensure only legally allowed combinations of components with compatible modification states are fitted using a graph database. These rules are easily expressible as data triples in linked data.
    - Linked data may be queried and surfed, and due to the underlying logic new relationships may be inferred and established, (e.g., if people are linked entities, then "my Mother's brother is my Uncle", the Uncle relationship is inferred from rules).

The advantage of URI/URLs is that it is both a unique identifier as well as an index pointing to other related data that can be accessed directly by computers. Might this capability make using part and serial numbers for components obsolete? This has potential for enormous cost savings in configuration management.

The Advent of the Citizen Data-Scientist

One of the biggest realizations in the author's career in terms of tools available to use for managing, analyzing, and visualizing insights from data was working in Ox Mountain. As a start-up, my colleagues and I have had to become generalists doing whatever has been needed to advance the company.

The author had been used to PdM engineers using MATLAB [9] and compiling models (using a MATLAB to C language conversion before C compilation) in Rolls-Royce, wrapping these in production systems. The costs for MATLAB and its compiler was too expensive for a start-up company, and so we have prototyped using Python [10] and used Java [11] as our production code.

One of the areas that is core to our business is data management and analysis. There has been a quiet revolution going on in the development of free, free-to-exploit, easy to learn/use, open source software tools. In the main there are open-source Big Data applications with the Apache software foundation, a set of programming languages and scalable middleware that can be used for free, and cloud services that may be experimented with very cheaply.

The Apache Software

The Apache software foundation has over 350 software projects and initiatives being actively undertaken. Many of these projects create large server and scalable technologies for Big Data. The list is too large to discuss in the scope of this book, but those wanting to understand how important the technical development is should see for themselves [12].

Python

The Python programming language [13] has been recognized as a leading language for data science and predictive technologies. It also has over 100,000 software libraries that may be downloaded and used including those that are at the forefront of current Machine Learning and Artificial Intelligence research and development.

Python is designed to be user friendly and easy to learn and is known to be productive in terms of the ease programmers find to build solutions. The language is interpreted and execution speed is relatively slow, but it is faster than interpreted MATLAB.

There are a number of software libraries associated with Python that have been written and compiled in the C programming language, compensating for some of the speed shortfall. These include Numpy [14] for fast mathematics and Pandas [15] providing a DataFrame that enables rapid numeric computation and indexing. The DataFrame is better than an excel spreadsheet and is used in the Python sci-kit learn libraries that has an extensive set of machine learning algorithms, along with simple-to-follow training and examples. Python graphics are included in a library called Matplotlib [17] which will be familiar to persons coming to Python via MATLAB.

There are many other powerful and easy-to-use libraries in the Python ecosystem that are worth exploring including: Simpy or Salabim are Discrete Event Monte Carlo simulation libraries, useful for building RAM analysis simulations [21, 22].

There are micro versions of Python available for microprocessor programming, which may help development of IIoT prototypes.

R is a specialist programming language, designed to be used by statisticians, that complements Python. The syntax of the language probably fits a statistician's mindset, but engineers seem to prefer Python. R is free with unrestricted deployment licenses [18].

These tools can easily be learned and used by any person with the motivation to learn. There is a myriad of instructional videos on YouTube, programming support on Stack-overflow [19] and many organizations who will support and describe how best to exploit the analytics.

Other Open-Source Capabilities

NASA has released it GSAP [20] (Generic Software for Prognostics) tools as an open source suite of C++ (C++ is a programming language) that can be used and extended by anyone. GSAP provides an extensible environment where third-party prognostic models can be embedded and run.

Key Take-Away Points

1. PdM is still emergent as a technology even without other new technologies discussed in this chapter that will also accelerate PdM's capability.
2. Being able to apply PdM to gain maintenance credits so that PdM becomes mainstream in aerospace maintenance will be a next major step to increase yield.

3. The emergence of Industrial IoT will transform PdM in making sensing data far cheaper and ubiquitous compared to today's technology.
   a. IoT needs to mature in two aspects: security and commoditized with industrial packaging before its widespread adoption in industry.
4. Tools for data management and analysis are available for free, with low-learning thresholds that allow easier adaptation and adoption. Support costs are still applicable.
   a. The development of major new data-management capabilities is open source.
5. The open source tools Python and R have very powerful libraries that are easy to use, and at the leading edge of Artificial Intelligence and Machine Learning.
6. Time series data analysis for PdM is very different from that commonly applied to financial (stock market) analysis.

References

1. IBM., The Big Date Four vs., https://www.ibmbigdatahub.com/infographic/four-vs-Big Data, accessed January 2019.
2. Wikipedia Contributors, "Apache Hadoop," https://en.wikipedia.org/wiki/Apache_Hadoop, accessed January 2019.
3. Wikipedia Contributors, "Map Reduce," https://en.wikipedia.org/wiki/MapReduce, accessed January 2019.
4. Wikipedia Contributors, "Industrial Internet of Things," https://en.wikipedia.org/wiki/Industrial_internet_of_things, accessed February 2019.
5. Wikipedia Contributors, "5G," https://en.wikipedia.org/wiki/5G, accessed February 2019.
6. Wikipedia Contributors, "Industry 4.0," https://en.wikipedia.org/wiki/Industry_4.0, accessed March 2019.
7. Wikipedia Contributors, "Aircraft Maintenance Checks," https://en.wikipedia.org/wiki/Aircraft_maintenance_checks, accessed January 2019.
8. Wikipedia Contributors, "Internet Protocol Version 6 (IPv6)," https://en.wikipedia.org/wiki/IPv6, accessed November 2018.
9. Wikipedia Contributors, "MATHLAB," https://en.wikipedia.org/wiki/MATLAB, accessed January 2019.
10. Wikipedia Contributors, "The Python Programming Language," https://en.wikipedia.org/wiki/Python_(programming_language), accessed April 2019.
11. Wikipedia Contributors, "The Java Programming Language," https://en.wikipedia.org/wiki/Java_(programming_language), accessed April 2019.
12. The Apache Foundation, "Apache Projects," https://projects.apache.org/, accessed January 2019.
13. The Python Organisation, "Python Language," https://www.python.org/, accessed January 2019.
14. The Numpy Project Organisation, "Numpy Python Library," http://www.numpy.org/, accessed January 2019.
15. The Pandas Project Organisation, "Pandas Python Library," https://pandas.pydata.org/, accessed January 2019.

16. The SciKit Learn Project Organisation, “SciKit Learn Python Library,” https://scikit-learn.org/stable/, accessed January 2019.

17. The MatPlotLib Project Organisation, “MatPlotLib Python Library,” https://matplotlib.org/, accessed January 2019.

18. The R Organisation, “The R Programming Language,” https://www.r-project.org/, accessed December 2018.

19. Stackoverflow Contributors, “Stackoverflow Questions and Answers,” https://stackoverflow.com/, accessed June 2019.

20. NASA, “NASA GSAP Prognostics Library,” https://ti.arc.nasa.gov/news/GSAP-Release/, accessed August 2019.

21. The Simpy Project Organisation, “Simpy Python Library,” https://simpy.readthedocs.io/en/latest/index.html, accessed August 2019.

22. The Salabim Project Organisation, “Salabim Python Library,” https://www.salabim.org/, accessed August 2019.

CHAPTER 11

A Summary, Future States, and Things to Look For

Knowledge about PdM, how it works, how it may be successfully applied and sustained is immature. Systems are being sold and adopted that are underdeveloped and will only add marginal value. Indeed, some immature systems will merely add overhead cost to an existing maintenance regime.

PdM is still evolving technically and procedurally. It is on the cusp of being justified in aerospace in taking maintenance credits and significantly expanding the scope of application. Along with PdM's development, the demands of maintenance in general are growing as societal and customer demands grow. More is expected, quicker for less cost and effort.

Emergent technology will further transform PdM because it will imply far more and cheaper sensing producing big volumes of new data, along with the tools available to interpret that data. Data has sometimes been cited as the new oil of business: it is certainly a resource that has much latent value ready to be released.

How Do You Start Implementing PdM?

The following bullet points provide a template for introducing PdM without an initial strategic-level investment, aimed to deliver rapid benefits:

- Start small, pick a failure that has high impact, with sensor data that PdM can solve by applying the questions and pre-requisite questions in this book.
- Set up a six-week duration project with innovative people and solve the problem using PdM principles.

- Show how the failure mode may be diagnosed in each failure event and show how prognostics can predict remaining useful life, allowing corrective maintenance to be scheduled that avoids the original disruption. There are plenty of free tools to use to build the analytics, but even if your IT is locked down in your organization you can still use Excel. Demonstration of a live project with real data and considerable value immediately establishes excitement and credibility to commit to a larger pilot project.
- Build on success and pick the next set of problems to solve. As you develop, you should take the time to reflect and review with the project team, learn lessons, and apply them going forward.
- As you pick up experience and knowledge, you can think about scaling the solution, and perhaps buying ready-built tools—your experience will inform you about your requirements as well as the advice gleaned from this book.

At a point where the experimenting and prototyping have proven value, the project should then switch to the assets (or systems) that drive the most cost in production. Apply the RCM process, but run it in a way that ensures PdM is fully considered and adopted.

Traditionally maintenance has been assumed to be a necessary cost for an asset-rich organization. This attitude is incorrect: PdM and the clever application of maintenance can actually increase revenue earned from assets. Consider examples where both reliability and availability are increased for assets: can this be exploited by further profitable utilization, as well as saving sunk capital in reducing spares and standby assets (reducing "static" inventory)? Each situation needs to be considered in context. For example, there is no profitable utilization to be gained by running all buses in a fleet at 3 am in the morning.

Maintenance can also address low-asset performance and efficiency, including fuel saving and reducing emissions that are becoming mandatory with new legislation. What about maintenance improving customer experience, if they use your assets (aircraft passengers)? How can noise, temperature, humidity, and comfort be enhanced as well as the elimination of unplanned breakdown?

Focus on reducing or eliminating events such as delayed departures, aircraft turn-arounds due to equipment failure. Cleanliness and appearance are important and engender trust in customers that clean assets are well maintained, and in their minds are preferred for their use. In a production line as part of an Industry 4.0 initiative, PdM can monitor product quality, delivering as a by-product production-line machinery health management output.

The Baby Boomers generation is retiring, taking a great deal of knowledge with them. The new generation of engineers are no less capable, but there are fewer of them. Apprenticeships that produced the last generations of technicians are no longer as common, or some of the quality of apprenticeships has declined. The degree of automation in PdM is considerable (if properly implemented), which compensates for this demographic trough. The newer generation of engineers needs to be more data and predictive-technology savvy compared to their predecessors, so they can take advantage and be far more productive.

The author once conducted a consultancy for a large utilities company, where drastic manpower reductions had taken place in a fossil-fuel fired power station. The residual crew had the knowledge to run the power plant under normal running conditions but were unable to cope with abnormal operations. After a planned outage, they conducted a plant start up that went drastically wrong, they boiled dry and wrote off a complete boiler that resulted in costs far exceeding the savings made from the manpower reductions. Making such drastic cuts based on economic drivers and not recognizing the true value of knowledgeable people may be counterproductive. Systems thinking and understanding that major influencing feedback loops (causes and effects) often have long delays would be a useful skill in these circumstances.

Management needs to be accustomed to thinking and making decisions embracing probabilistic data and uncertainty.

PdM is still emerging and knowledge about it is still ambiguous. There are many new suppliers who are trying to generate an edge, some unfortunately by making claims that do not stand up to detailed scrutiny. This book attempts to explain some of this hype and improve understanding. Hype and inflated expectations will continue to reduce until PdM becomes more mainstream, and high-quality PdM suppliers emerge.

PdM Analogies

PdM principles are often understood better if we use examples that most people experience and can appreciate. One of the best examples to use is the car.

PdM Triggers Restorative Action. An example may include sensing tyre pressure, where an alarm is set off when tyre pressure falls too low, and the pressure needs to be restored to a nominal normal value.

As an added sophistication, the PdM system could also sense ambient temperature, and adjust the alarm or alert levels because it may be more desirable to lower tyre pressures in winter conditions to try and increase traction. The tyre may become warmer after the car has run for a few miles compared to the temperatures after a car has been parked for a few hours. Temperature will influence pressure, and so the pressure reading may be normalized by compensating for tyre and ambient temperature.

PdM Triggers Replacement Action. Two examples may be used here.

- **One or more of the rear braking light clusters have failed**. This event occurring in the UK warrants a stop by traffic police, the police would caution the driver to rectify the problem at the earliest opportunity. How would a driver know that the braking lights had failed, when driving the car? It is very difficult for the driver by themselves to observe the lighting working when they press the brake pedal. The consequences of this failure might be catastrophic if another event occurred. The other event could be you are driving with a lorry following you. You observe a hazard in the road that you need to brake for. Because of the partial loss of braking lights in your car, the lorry driver fails to notice you are braking and crashes your rear. This situation is known in the Reliability Centered Maintenance (RCM) framework as a "hidden failure". The device has failed but the failed state is unnoticed by the operator, undertaking their normal duties (driving the car). A second event (lorry crash) occurs where the impact and consequences are far greater than the first failure. How can PdM help? What if all cars had a fiber-glass strand (or set of strands) that ran from the rear lights to the car dashboard? The driver could start the car, test the brake, and observe the lights working through the fiber on their dashboard (This would be the author's preferred solution). Alternatively, PdM could have a light sensor fitted, that worked in conjunction with brake position (brake applied or not applied), to trigger an alert if the lights did not turn on when required. With a bit more sophistication, during the car starting sequence the brake lights could be tested and reported.
- **The Pollen/Air Filter Is Blocked**. Blocked-filter detection is quite common, but in the author's experience if it is not implemented correctly can lead to false positives. If the filter is blocked, the air conditioning or heating may be reduced in its effectiveness leading to passenger discomfort. The author suffers from hay fever and opening a window for forced ventilation is not particularly satisfactory when

ambient pollen levels are high. A blocked filter might be detected by measuring the difference in pressure across the filter, with the fan speed setting being used to regulate the alert function. The more blocked the filter the higher the differential pressure (most engineers use the term "DP" for differential pressure), proportional to the speed of the fan. An alert may be sent if DP rises above a threshold warning the driver to service the car and replace the filters for new to restore the function. However, there may be a problem with this arrangement. What if downstream of the filter the pipework carrying the air came adrift, reducing the back pressure behind the filter.? The physical effects are that the flow of air would increase, and the DP would also increase setting off the blocked filter alert. In this situation the alert would be wrong (a false positive) the filters are not blocked. A more sophisticated PdM system that would eliminate this false positive is fitting a sensor that measured airflow, as well and another sensor measuring DP. The accurate diagnosis of a blocked filter is definitive when DP increases and flow decreases. This is an example illustrating the thinking that needs to occur using engineering-domain expertise to properly specify accurate PdM systems. The arrangement of sensing flow and DP would also widen the scope of diagnosis, because besides monitoring the filters, the fan blade performance (losing aerodynamic performance) through getting dirty could be isolated as well as leaks in the ventilation tubing. By eliminating faults in these three areas, it may also be indicative if the car cabin environment cannot be maintained, that faults exist in the air conditioning or heating systems. For example, perhaps the air conditioning compressor needs re-gassing? The effectiveness of the PdM system is significantly improved.

Some may critique this, by saying that fitting extra sensors may be expensive, and make the whole sensing system more complicated, and therefore more unreliable. However, this valid point may be countered with the emergence of the Industrial Internet of Things (IIoT), sensors become smaller, cheaper, and far more reliable. The IIoT will not require the sensors to be cabled for power or for transmitting data, they will work with low-energy wireless and scavenge their own power. This slashes the cost of instrumenting by orders of magnitude compared with today's technology. This also implies the proliferation of sensors that enables a transformation in the coverage and effectiveness of PdM.

Key Take-Away Points

1. PdM is maturing within the current state of technology but will be transformed with the emergence of new technology. Embracing it now, positions organizations to take advantage of the future transformation.
2. The DNA of PdM is automation but keeping humans in the decision-making loop. This improvement of productivity will compensate for the loss of knowledge with Baby Boomers retiring. Much of their knowledge needs to be captured in PdM systems. The automation should aim to make the human-domain experts orders of magnitude more effective and productive.
3. Automation of the more mundane aspects of analysis are laudable, but never forget that the important components of a system are the people. Automation should make them more productive and improve their work experience by relieving them of some of the more mundane parts of their jobs.
4. Start PdM as a small project, with a painful failure mode with lots of events and data. Prove value, establish credibility, and build knowledge. Then scale up.
5. RCM will be extremely valuable to apply once you know how to operate PdM.

CHAPTER 12

An Example PdM Case Study Using Open-Source Development Tools

This chapter will outline a simple PdM use case and show how time series data may be analyzed and processed to detect anomalous behavior. This is intended to be a taster, the use case will not include diagnostics, prognostics, or a managed alerting system, because they would be unfeasible within the constraints of this chapter. The author may develop a GitHub repository with other simple PdM use cases separate from this book.

The data will be simulated and the Python language within a Jupyter Notebook will demonstrate some of the concepts discussed in this book.

The case study will use a single variate Kalman Filter, that has been developed by Dr. Michael Provost outlined in [1]; Chapter 6.

The Python script for the univariate Kalman Filter is shown below, this will allow the script to be reproduced in reader's own systems so they can experiment with the Kalman Filter algorithm:

```
import numpy as np
import pandas as pd


def univariate_kalman(values: any, alpha=0.2, beta=0.01, limit: float = 2.0)-> pd.DataFrame:
    """
    outputs a dataframe with columns for input data (observations) Time, Trend, Delta and smoothed estimates

    :param values: Input raw data as a single dimensional ndarray of floats
    :param alpha: enables the smoothing effect of the Kalman filter to be tuned
    :param beta: with alpha enables responses to the underlying trend of the data to be tuned
    :param limit: a value that designates a step change or outlier boundary
    :return: A DataFrame with input data, intermediate results and output smoothed results
    :raises ValueError: if alpha or beta are not in the range {0 .. 1}
    """
    # alternative to setting beta, use beta = alpha ** 2 / ((2-alpha) * delta_t) if wanted
    # delta_t is the average time units between observations (if uniform this value is not required)
    if 0 < alpha > 1.0 or 0 < beta > 1.0:
        raise ValueError(f'Both alpha value: {alpha} and beta value: {beta} must be between 0 and 1')
    df = pd.DataFrame(values, columns=['Observations'])
    # add and fill in new columns with default values, ensure numeric types are float
    new_columns_and_defaults = {'Time': 1.0, 'Trend': 0., 'Delta': 0., 'Estimate': 0}
    df = df.assign(**new_columns_and_defaults)
    # this makes sure estimate[0] is set to the observation as the initial condition for the kalman calculations.
    df.loc[0, 'Estimate'] = df.loc[0, 'Observations']
    # do the Kalman calculations-deliberately start from index 1-uses a for loop for clarity
    for kalman_index in range(1, len(df)):
        df['Delta'][kalman_index] = df['Observations'][kalman_index]-(df['Estimate'][kalman_index-1] +
                                                                        df['Time'][kalman_index] *
                                                                        df['Trend'][kalman_index-1])
        # see if Kalman must ignore outlier, or snap to new trend after a genuine step change
        if abs(df['Delta'][kalman_index]) > limit:
            b = df['Delta'][kalman_index-1]
            a = df['Delta'][kalman_index]
            # check is previous above limit and both deltas share the same sign
            if b > limit and abs(a+b) == abs(a) + abs(b):
                # reset is required
                df['Estimate'][kalman_index-1] = df['observations'][kalman_index-1]
                df['Trend'][kalman_index-1] = 0.01
                df['Delta'][kalman_index] = df['Observations'][kalman_index] + 0.01
                continue
            else:
                df['Delta'] = 0
        df['Estimate'][kalman_index] = (df['Estimate'][kalman_index-1] + df['Time'][kalman_index] *
                                        df['Trend'][kalman_index-1]) + alpha * df['Delta'][kalman_index]
        df['Trend'][kalman_index] = df['Trend'][kalman_index-1] + beta * df['Delta'][kalman_index]
    return df
```

Univariate enhanced Kalman Filter, from algorithms written by Dr Provost. Copyright for the code CE Dibsdale (2017). Use of this code is at your own risk and if used you should recognize and acknowledge the authors.

Plots

The first set of plots is a comparison of the Kalman Filter algorithm compared with an Exponentially Weighted Moving Average (EWMA) and a Moving Average (MA) plot.

The data is generated from a "random walk" algorithm with added noise to simulate a time-series plot, that may be typical of transient behavior in an asset. The data consists of 100 points. The EWMA and MA algorithms have a moving window (from which the averages are calculated) of ten historical data points. The plotting is done with the Python MatPlotLib library. The blue trend line is the random raw data and the red trend lines are the estimated smoothed lines.

You may notice the Kalman Filter plot in Figure 12.1 displays less "lag" and tracks the raw data more closely. Both the Kalman Filter and the EWMA plots may also be tuned to be more responsive (more or less smoothing) by adjusting the alpha and beta values in the Kalman Filter algorithm. It is also possible to change the moving windows to calculate the means using the MA algorithms and also to adjust the degree of exponential in the EWMA algorithm.

FIGURE 12.1 Comparison of time-series smoothing algorithms.

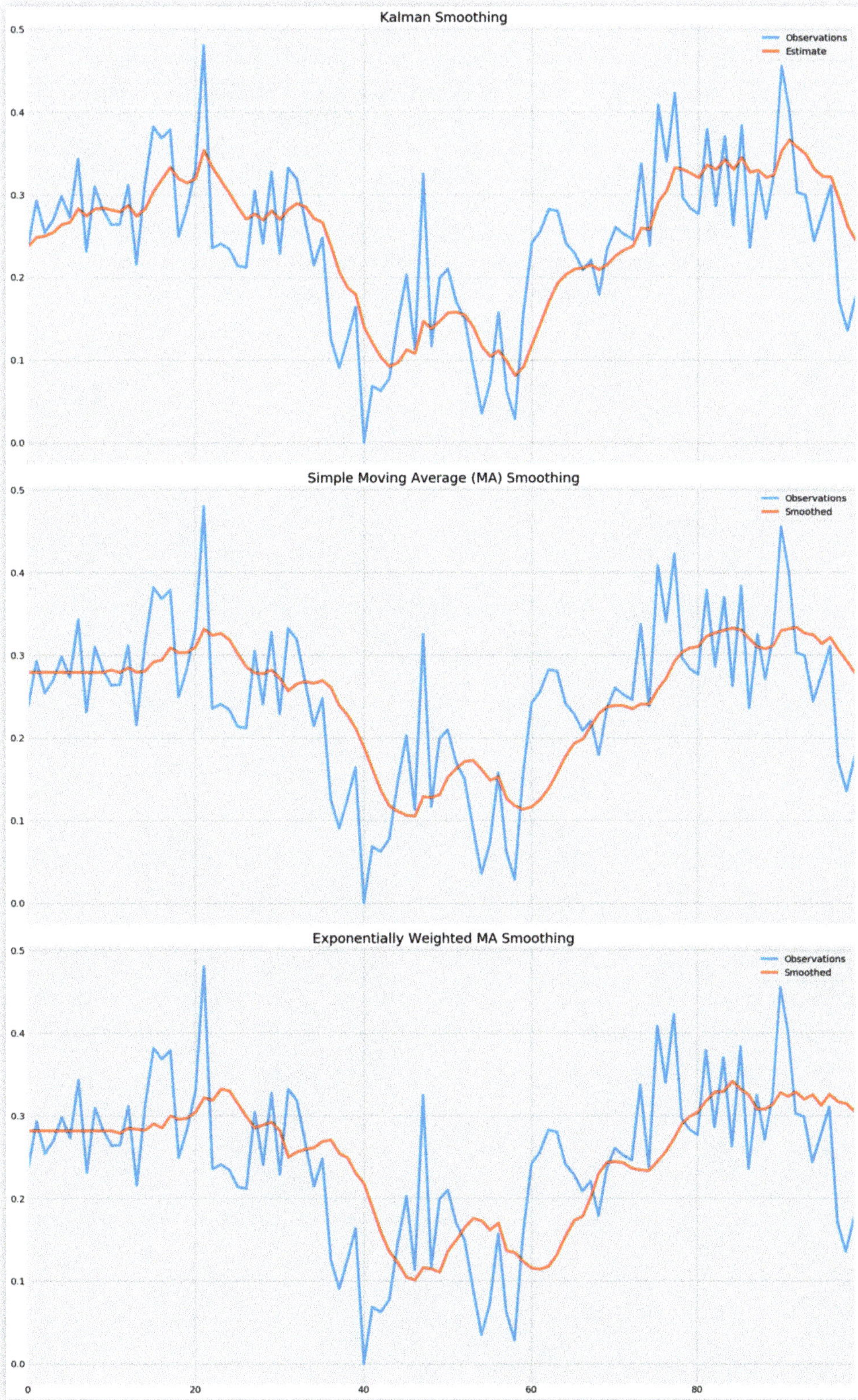

The next set of charts shows two time series plots with the upper plot showing a trend with a step change about halfway through. The lower plot shows an extra output variable inside the Kalman Filter algorithm that shows the rate of change in the underlying trend. It is this plot that could be set with a threshold detection (at a level of 0.06 in the lower plot). This would set an alarm when a significant step occurs that indicates that something fundamental has occurred that may be indicative of failure. Step changes may also be indicative of restorative maintenance being done, where the performance of machinery has improved. The diagnostic function is required to differentiate between failure and other events to ensure false positive are minimized. The step change may not be big enough to set off a threshold alarm in the smoothed trend in the upper chart, but it is enough to set off a "step change anomaly detector" (say set at 5.5 in the upper chart) that can then be supplied into a diagnostic function (Figure 12.2).

FIGURE 12.2 Kalman Filter step change novelty detector, using the Kalman trend parameter.

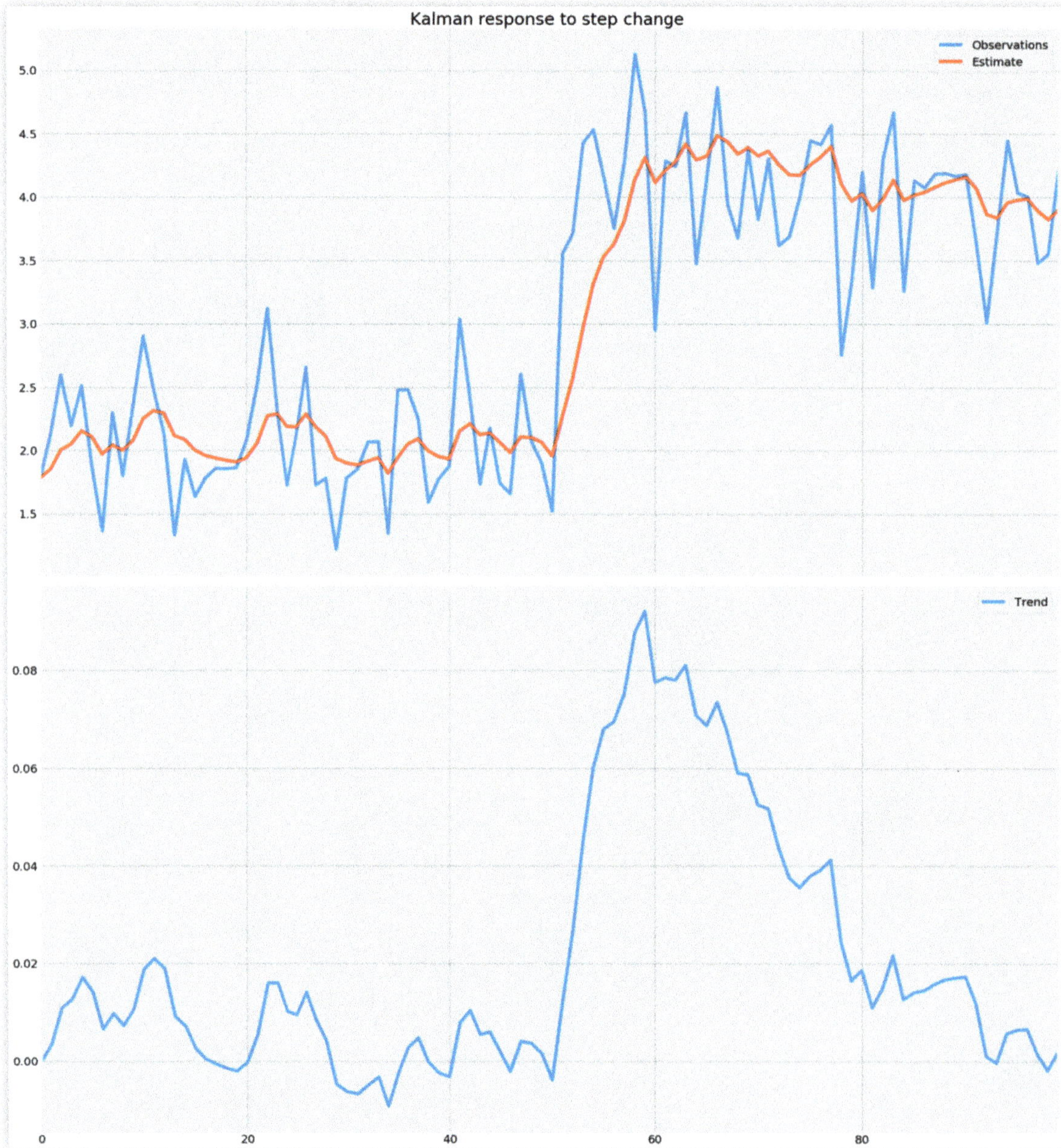

The third example for a Kalman filter below in Figure 12.3 is when a trend is ramping up slowly over a period of time. A simple threshold on the raw parameter may only trigger an alert after a considerable time after the ramp increase has started. It would be better to detect the ramp increase anomaly earlier. This, coupled with other anomalies, may trigger an earlier diagnosis allowing a longer P-F period and calculated Remaining Useful Life.

FIGURE 12.3 Kalman filter ramp change novelty detector.

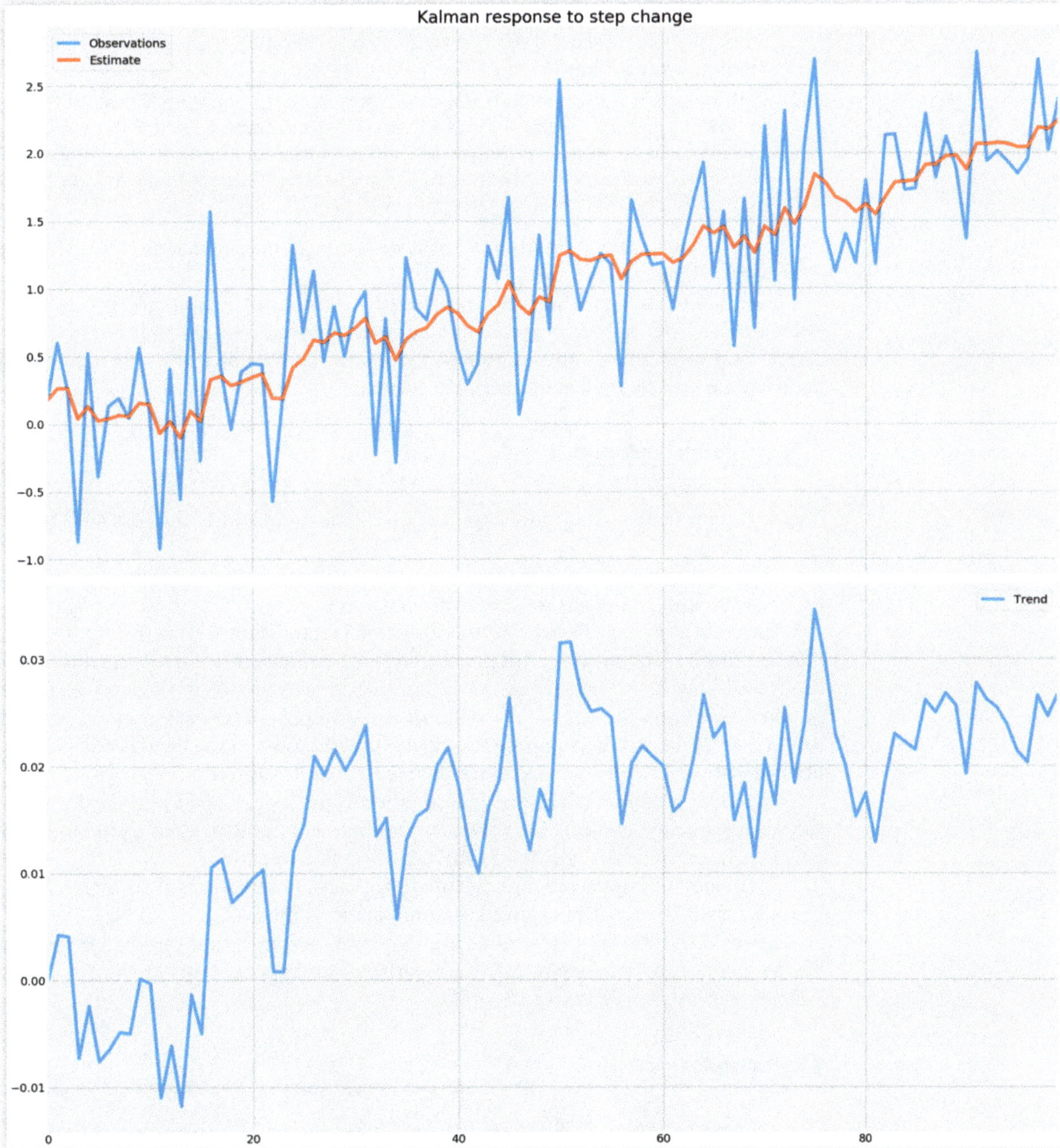

There are a few possibilities here to construct a Novelty detector.

1. By taking the estimated (smoothed) trend from the top chart in Figure 12.1 and calculating the slope between points.
2. By taking the rate of change trend line from the bottom chart in Figure 12.3 and setting a threshold (say at 0.015) and counting an arbitrary number of consecutive data points (say 5) that all stay above the threshold.

Summary Notes

This chapter has shown the power of Kalman Filtering to

1. Smooth noisy timeseries data more effectively than the simpler moving-average algorithms.
2. Output rate of change and acceleration. Experience in applying Kalman Filters to PdM suggests acceleration is rarely needed, it is more applicable to problems such as future position prediction for moving target in gunnery. These may be used as anomaly detectors to detect ramp or step changes that captures the earliest evidence of changing machinery behavior.

These demonstrate a big improvement in having simple threshold detection and anomaly detectors alone.

There are other algorithms that may be applied including Auto-regression Moving Averages (ARMA) or Auto-regression Integrated Moving Average (ARIMA) that have similar outcomes. There are many simple-to-follow tutorials and YouTube videos that demonstrate how to implement these algorithms.

Code Notes

These notes provide some more explanation of the code and with what and how it was written.

The code was written to enhance understanding and mirrors the algorithm written by Dr M Provost [1] and acknowledges his original work.

The code is compatible with Python version 3.6 (and above) with the Numpy and Pandas libraries (any version), that provide advanced numerical routines on vectors and the DataFrame (the df object in the code), that makes Maths on indexed arrays of numbers very easy. The code could be considerably improved for performance, but this would detract from understanding the algorithm. Python and the Pandas and Numpy libraries can be downloaded and exploited commercially for free.

The code is commented and includes Python Type labeling and a python-doc entry (between the triple quotes). The type annotation is not mandatory in Python (as the language is not statically typed) and will not affect the execution.

The values may be passed into the function as a list (a vector) of values representing readings of time series data sampled at a uniform time interval.

The df (DataFrame) return value is a Pandas data frame, that can then be passed into a separate function for plotting. The algorithms for Exponentially Moving Averages and plotting are not reproduced here.

Reference

1. Dr. Michael. Provost, "Servitization and Physical Asset Management," SAE, 2018.

glossary

Term	Description
Big Data (Including the 4 x V's description)	Is a catch-all term used to describe data processing technology to economically capture process and store very large volumes of a wide variety of data (including unstructured data). The technology has been developed as open source software, and reduces costs significantly from traditional database and data warehousing technologies. The four Vs provide a descriptive framework to describe the attributes of Big Data. **Volume:** describes the accelerating global production of data, that is doubling every 2 years. The volume is such that the older generation of database management systems are unable to economically scale to contain them. **Variety:** describes the variety of the types of data being produced image, video, audio data is increasing and the application of AI and Natural Language Processing (NLP) is starting to make these varieties of data classifiable and searchable. Video image data used in augmented reality may greatly improve maintainability. **Veracity:** describes the truthfulness of data, and whether it can be trusted. On one hand, using statistics may help filter out erroneous data, and automatically deal with lower quality data. But on the other hand, data selection may include bias that will be replicated in AI systems. Much work on Ethics in AI need to be done to improve veracity. **Velocity:** There are many use cases where the speed of systems must approach real time. Google helped develop Big Data because their search engine needs to return credible search results often in under a second so that user experience is optimal. PdM may need near real time if actions need to be taken if P-F intervals are short. This is best done on the asset, where the operators are optimally positioned to take remedial action. Real time is not needed for events with longer P-F intervals, where remedial action might be best initiated remotely and centrally. PdM does not necessarily need real time but has to be timely.
Black Swan Event	Is a metaphor using a medieval saying that due to evidence the likelihood that black swans existed (in the then known world) was so unlikely it was almost certain not to exist? Then black swans were discovered with the exploration of Australia. The saying is now representative of a very unlikely event that is still possible.
Capta	Is a new term in the amended DIKW model presented in this book, that captures the ability of computing machines to produce or infer finite sets of data (facts) that have relevance in a context. Capta is relevant subsets of data that a human can synthesize information from, using the human ability to sensemake.
Condition Based Maintenance (CBM)	Condition-based maintenance, or on-condition is a type of maintenance that determines the condition and remaining useful life in order to maximize the utilization of economic life of machinery. If the condition has degraded sufficiently, remedial or corrective action may be planned and resources pre-disposed so recovery is as least disruptive as possible. The US armed forces (DoD) also have a concept of CBM+, where they exploit RCM to specify as much CBM as possible (including PdM) and institute a continuous improvement program to increase effectiveness and efficiency of the CBM, but also to widen the scope of applied CBM as possible. The CBM+ concept also incorporates integral prognostics. See the handbook at https://www.dau.edu/guidebooks/Shared%20Documents/Condition%20Based%20Maintenance%20Plus%20(CBM+)%20Guidebook.pdf
Discrete event simulation	Is a means of simulating systems where objects change state over time. The events may be triggered by a randomly sampled period of time. The simulation may consist of queues, objects, and services that interact via encoded rules. This technology is used in RAM and through-life cost calculations.

Term	Description
Failure mechanism	This is where basic physical processes in thermodynamics, chemical, electrolytic and galvanic action, stress, cyclic fatigue, tribology and impact/loading-distortion domains that when applied lead physical assets to functionally fail. Failure mechanisms can act alone or interactively to cause failure. Failure mechanisms may also be more or less likely to act depending on operating environments. For example, corrosion is more likely in salt-laden moisture environments.
Failure Mode	Is a description of the state of an asset when it is failed. This is not the same as a failure mechanism. A failure mechanism leads to failure modes. A failure mode example is "unacceptable leakage" that may be caused by corrosion, erosion, impact deformation or cyclic fatigue. PdM is configured to monitor the onset and degradation of failure modes. It is possible for PdM to measure the local effects of failure modes or the underlying failure mechanisms itself.
Industry 4.0	Is an amalgamation of many emergent technologies to build data-driven configuration control and monitoring of production lines. The technology takes advantage of production line robotics and additive manufacturing to vastly reduce manning. This provides opportunities to manufacture locally on-demand Just in Time. This saves logistics costs. The instructions to manufacture may be globally transferred as data, meaning transportation of physical goods is reduced. As part of the Industry 4.0 mix it uses PdM as a major component of product quality assurance and machinery health.
(Industrial) Internet of things (IoT)	Is based on miniaturization of very low-powered sensors processing and memory for small devices that can link to the internet. Many of the smart devices communicate through localized mesh wireless networks. This offers the opportunity of fitting many more sensors without data or power cables cheaply, with local hubs on remote machines that can process data and take more autonomous decisions. This technology may transform the capability of PdM.
MSG3	Aerospace, Maintenance Steering Committee 3. The body and current version for guidance for the application of maintenance. MSG3 is derived from the same resource as RCM that is used in other industries, both from the Nowlan and Heap 1978 report. This guidance is used by aerospace companies as a set of baseline principles for designing and implementing aircraft maintenance regimes. Each regime is then formally approved by regulators.
NDT or NDE	Non-destructive testing or examination: techniques such as ultrasonic, magnetic eddy current, and dye-penetration tests used to discover cracking in metallic structures, where the tests do no harm to the structure. These tests and examinations belong to the family of condition monitoring maintenance. The results of the tests are traditionally assessed by qualified specialists who will report the condition of the inspected structure. These tests are types of on-condition maintenance usually used to monitor long-term degradation.
Predictive Maintenance (PdM)	Predictive maintenance is a way of exploiting digital control, data, communication, and processing, married to predictive technologies (including machine learning and artificial Intelligence) to automate condition monitoring. PdM usually uses fixed sensors and so monitoring is continuous and the speed of automatic analysis is only limited by processing delay and communication latency and bandwidth, enabling the remote management of failures with shorter P-F intervals than the previous generation of on-condition maintenance techniques.
Python	Is a scripting programming language that has been designed to be easy to learn with a syntax that reads closer to English than most other programming languages? Python has gained popularity because of its software libraries associated with Machine Learning and Artificial intelligence. It is free to procure and the resultant code is free to deploy commercially. Python is a suitable replacement of and complements engineering tools such as MatLab that some people may find prohibitively expensive.
RCM	Reliability Centered Maintenance is a framework for designing implementing and monitoring (with a process called "age exploration") a maintenance regime. RCM and MSG3 are closely aligned both being developed from the Nowlan and Heap (1978) RCM report. A regime or process claiming to be RCM must comply with SAE JP 1011. The compliance standard was introduced because of the emergence of hybrid or accelerated RCM processes that miss out some of the important stages.
Specificity and sensitivity	Are terms used with machine learning classifiers, confusion matrixes, and RoC curves. This is where classifiers may be tuned to be more sensitive, where the identification of true positives is enhanced (correctly identifying a failure), or more specific where the identification of true negatives is enhanced (correctly identifying where a failure is not present).

author bio

Charles E. Dibsdale joined the Royal Navy in 1974 as a technician apprentice (artificer) specialising in Marine, Electrical and Nuclear Engineering, and eventually operated and maintained nuclear propulsion and submarine systems in all levels of seniority up to supervisory plant and maintenance positions. During this period, Mr. Dibsdale was commended for two studies on the rationalization of nuclear safety checks with planned maintenance, and on reporting deficiencies in naval systems for equipment configuration management. Mr. Dibsdale gained an honours degree in Computer science in 1996.

Leaving the service after 23 years, the author joined Rolls-Royce Naval Marine as a maintenance reliability engineer supporting the UK submarine fleet. Among different accomplishments, Mr. Dibsdale won funding for and then project managed the build of a FRACAS system (Failure Recording and Corrective Action System), that improved defect-reporting and equipment-modification response time and value by an order of magnitude. This system was later used by the UK Ministry of Defence (MoD) to specify the reliability requirements for the (then) new Astute Submarine Class build contracts.

In 1999, a new opportunity opened up to transfer to a new Rolls-Royce and SAIC joint venture. He became part of the team that developed the predictive maintenance capability that enables Rolls-Royce's TotalCare™ service offerings and grew to be the global capability owner for predictive maintenance. As a co-inventor of a patented predictive maintenance diagnostic method, Charles Dibsdale also engaged with standards groups, helping write engineering standards for risk-based maintenance in oil and gas (API 691) and looking at ways of predictive maintenance gaining maintenance credits in aerospace. He has contributed to six textbooks for Integrated Vehicle Health Management (IVHM) and Through-life Engineering Systems. Mr. Dibsdale received a Master's degree in Information Systems in 2006.

After finishing his career at Rolls-Royce in 2015, he co-founded a new company, Ox Mountain Ltd. Today Charles Dibsdale is an engineering domain expert guiding the development of software delivered as a service that automates and optimises engineering processes, including maintenance management and reliability engineering. Charles continues his work in research and prototyping engineering, so new software service solutions can be offered in the marketplaces that rely on complex machinery to operate effectively and efficiently. Charles has continued providing lectures on predictive technologies and maintenance management to a number of university advanced courses in the UK and Australia.

www.ingramcontent.com/pod-product-compliance
Lightning Source LLC
LaVergne TN
LVHW070910210926
848482LV00017B/1044